Getting Started with Grafana

Real-Time Dashboards for IT and Business Operations

Ronald McCollam

Apress®

Getting Started with Grafana: Real-Time Dashboards for IT and Business Operations

Ronald McCollam
Somerville, MA, USA

ISBN-13 (pbk): 978-1-4842-8308-0   ISBN-13 (electronic): 978-1-4842-8309-7
https://doi.org/10.1007/978-1-4842-8309-7

Managing Director, Apress Media LLC: Welmoed Spahr
Acquisitions Editor: Jonathan Gennick
Development Editor: Laura Berendson
Coordinating Editor: Gryffin Winkler

Cover image designed by Freepik (www.freepik.com)

Distributed to the book trade worldwide by Springer Science+Business Media LLC, 1 New York Plaza, Suite 4600, New York, NY 10004. Phone 1-800-SPRINGER, fax (201) 348-4505, e-mail orders-ny@springer-sbm.com, or visit www.springeronline.com. Apress Media, LLC is a California LLC and the sole member (owner) is Springer Science + Business Media Finance Inc (SSBM Finance Inc). SSBM Finance Inc is a **Delaware** corporation.

For information on translations, please e-mail booktranslations@springernature.com; for reprint, paperback, or audio rights, please e-mail bookpermissions@springernature.com.

Apress titles may be purchased in bulk for academic, corporate, or promotional use. eBook versions and licenses are also available for most titles. For more information, reference our Print and eBook Bulk Sales web page at http://www.apress.com/bulk-sales.

Any source code or other supplementary material referenced by the author in this book is available to readers on GitHub.

Printed on acid-free paper

To my partner, my friends, and family (both birth and chosen), who put up with my absence for far too many evenings and weekends. Thank you!

Table of Contents

About the Author

Ronald McCollam is a "geek of all trades" with experience ranging from full stack development to IT operations management. He has a strong background in open source software dating back to when a stack of 3.5" Slackware floppies was the *easy* way to install Linux. When not on the road for work or in his lab building robots that can operate a Ouija board, Ronald resides on his back porch in Somerville, MA, with a frosty beverage in hand.

About the Technical Reviewer

 **Alfons Muñoz** is a Program Manager at C2C Global, the Google Cloud Customer Community, where he specializes in developing a vibrant cloud community. He is 3x Google Cloud certified, including Associate Cloud Engineer, Cloud Architect, and Collaboration Engineer, and is also a Google Cloud Partner. As a big fan of data-driven companies, Alfons is always involved in the data management projects in the company and enjoys moderating events in the community where he interviews cloud and data specialists.

Acknowledgments

Writing a book turns out to be a lot of work. That's something that everyone says, so I was prepared going into this to put my head down for as long as it took. What I wasn't prepared for was the impact it would have on everyone around me.

To the team at Apress, your help and advice has been fantastic. Jonathan Gennick, for kicking off the process and helping me understand just what is needed; Jill Balzano, for cat herding and keeping on top of the process; and especially Alfons Muñoz, for reviewing every piece of technical instruction and checking that every step worked and every screenshot was clear. (If I had realized how tough your job was, I might have left out the installation instructions for the Raspberry Pi at least!)

To my team at Grafana Labs who put up with me bleary-eyed and exhausted every Monday morning after working through the weekend, and especially Eldin Nikocevic who helped review and hone my muddled ideas into something approaching sense. If any of you want to tackle a book next, I'll be there to support you if I can't talk you out of it.

For every relationship I ignored or neglected, you have my thanks and apologies. To Silver Hawthorn for understanding emphatically how exhausting this process is – it meant a lot to hear "I know!" Most especially to Dan Nicholas and Rowan McVey; I flaked out on you so many times that it's a wonder you still talk to me. Thank you.

To my parents, Ron and Mavis McCollam, for all of their support and encouragement through my life. Dad, I wish you could have seen this published, but I know you were proud regardless.

Finally, to my partner, Andi McCollam, thank you for all your blocking and tackling of life events and social obligations for the months when my evenings and weekends disappeared into building environments and taking screenshots. I'll eventually walk the dog again, I promise.

Introduction

The first time I saw Grafana I fell in love.

When I started monitoring IT operations at a legal document analysis company 20 years ago, there weren't a lot of good options for simply visualizing what was really going on. There were plenty of tools to capture data from the bulky, noisy racks of hardware in the datacenter, but they were all limited to showing simple statistics in tables or (in the case of the really fancy ones) line graphs that could plot a single set of metrics over time. "Monitoring" meant staring at these tables and charts at least a few times per week to look for indications of trends that might become problems down the line. Automated alerts were simple thresholds for CPU or memory utilization that pointlessly woke me up at 3:00 in the morning, but still had to be checked and reset before I could go back to bed. I spent as much time trying to build ways to see and manage my data as I did in actually using it productively.

We've come a long way in the last couple of decades. Most services now run in cloud environments where individual servers no longer matter. At the same time, the move to microservice architectures means that the number of services that an operator or Site Reliability Engineer (SRE) cares about has exploded. The simple tools from my datacenter days are still there, but their singular focus can't give the broader context that we now need to understand how these complex components interrelate.

That's where Grafana comes in. Grafana is the gold standard for live, real-time data visualization. It works with your existing tools and talks to your data where it lives. Rather than ingesting static point-in-time CSV dumps, Grafana runs queries against databases or services. Forget exporting and importing, no need to copy or move data; Grafana sits on top of your existing infrastructure and just lets you *see*.

You might be a manager who needs to meet specific service-level objectives (SLOs) for your business, wondering how best to track the service-level indicators (SLIs) that tell you if you're meeting your goals. Perhaps, you've just read the Google SRE Handbook and want a good way to track the Four Golden Signals for your application. Or maybe you just care about data and want a beautiful, easy way to visualize it! This is the book to get you started.

You'll first learn how to get up and running with Grafana in the cloud and how to create simple dashboards. From there, you'll deploy Grafana in your own environment, connect it to your own data, and manage user access and permissions to that data.

Next, you'll see how to design beautiful and effective dashboards that show your data in the most meaningful way. You'll learn how to build powerful workflows that guide people to the most relevant information, even when that information is scattered across many tools and storage systems. Templates and variables will let you create comprehensive views of data with minimal effort, and alerts will warn you about problems that occur when you're busy elsewhere.

Finally, you'll explore the full power and extensibility of Grafana. You'll see how to scale Grafana and treat its configuration and dashboards as code. You'll use the Grafana API to automate processes and provision new environments. You'll even get a peek at some of the enhancements available in the Enterprise version of Grafana.

As "monitoring" has grown to "observability," we've moved from looking at simple metrics from a single program or server to thinking about systems. A high CPU utilization doesn't matter as long as transactions are being processed within the business's SLOs. A server going offline entirely is fine if the service heals itself and keeps running. I might care to know that it happened, but I certainly don't want it to wake me up in the middle of the night!

Grafana is everything that I wanted when I was (literally) keeping the lights on in a datacenter decades ago. It elegantly combines data from nearly any source into a simple and beautiful interface that can give both real-time views and historical context about anything you care to track. After discovering it, I threw away all of my home-built scripts and clunky tables and charts and have never looked back.

In this book, I hope to give you an introduction to the power that Grafana contains. The concepts and techniques covered here will enable you to build meaningful representations of your data that can be used to inform decisions about scaling, point out potential issues, and even show how seemingly separate systems interrelate.

Grafana is an amazing tool that can be used almost anywhere. I hope you will enjoy it as much as I do!

PART I

Getting Started

The fastest way to learn to use Grafana is to just start using it. In Chapter 1, we'll look at some ways to use Grafana, starting with the free tier of Grafana Cloud. This will give you a full Grafana environment without any need to install anything or mess around with setting up a system, virtual machine, or container. Instead, you can just create an account and immediately start building dashboards!

In Chapter 1, you'll learn how to create a Grafana Cloud account, how to log in to Grafana and navigate around the interface, and how to create and save a simple dashboard with some mock data.

Chapter 2 will show you how to work with panels to build and customize more complex dashboards. We'll also take a look at ways to experiment with and explore data quickly for those times when building a full dashboard isn't really needed.

This initial exposure to Grafana from the perspective of a user will lay the foundation for more advanced topics in Part II (Chapters 3–5), where we'll look at deploying and managing Grafana in various environments.

CHAPTER 1

Grafana Cloud

As a widely used open source tool, Grafana can be deployed in a nearly limitless number of ways and at scales ranging from a single instance on a pocket-sized Raspberry Pi up to highly available multiregion deployments with hundreds of nodes. Figuring out the best way to deploy it for your own environment can seem a bit daunting.

Fortunately, Grafana is also available as Software as a Service (SaaS), meaning that someone else has already done all the work to set it up correctly and make it available to you to sign up and use automatically. Since Grafana is open source, anyone can host a SaaS Grafana service. And because Grafana continues to grow in popularity, there are a growing number to choose from.

Note Deploying Grafana yourself isn't as hard as it might look! We'll walk through some common deployments in the next section. For now, starting with Grafana Cloud lets us get up and running quickly.

In Part I, we'll focus on Grafana Cloud, provided at $www.grafana.com$. Grafana Cloud has a few advantages for our purposes. For one, it's managed by Grafana Labs, the corporate backer of the Grafana project. So Grafana Cloud will always have the latest release of Grafana available, and since it's run by the people who make Grafana, you can count on it to be configured and patched correctly.

Another reason to use Grafana Cloud is that it provides a free tier that should provide plenty of resources to get you started with Grafana. (It even includes some other services such as metrics and logs storage so you can experiment with your own real data if you choose.) You can sign up without using a credit card or other forms of payment. And while you'll start on a "pro" tier that is time limited, the standard open source Grafana functionality will remain free after this initial trial expires.

© Ronald McCollam 2022
R. McCollam, *Getting Started with Grafana*, https://doi.org/10.1007/978-1-4842-8309-7_1

Creating a Grafana Cloud Account

To use the hosted version of Grafana, you'll first need to create a Grafana Cloud account. Start by opening your web browser and visiting the Grafana home page at *www.grafana.com*.

The web page may change from time to time, but look for a button that says something like "Create a free account." If you don't see anything like that, click the "Login" link in the upper right. You can then register for an account from the login page directly (Figure 1-1).

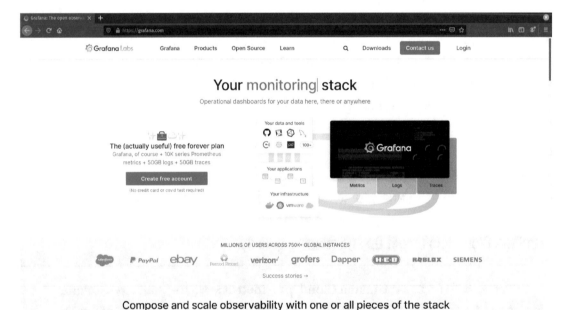

Figure 1-1. *The Grafana home page with the "Create a free account" button center left*

Next, you'll need to choose how you want to authenticate to Grafana Cloud. There are a number of options, including single sign-on (SSO) options. If you have an account on a supported service such as Google (or Gmail), GitHub, Microsoft, or Amazon, you can use that account to log in without having to remember any new credentials. Of course, the old standby of using your email address and creating a password is available too, as shown in Figure 1-2. Don't worry too much about which method you use – you can always add other accounts and authorization systems later.

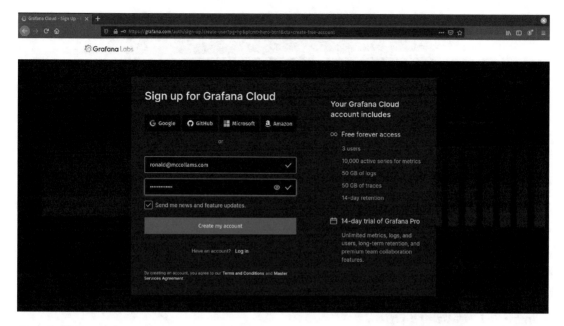

Figure 1-2. *Creating a Grafana Cloud account with an email address*

If you do choose to use an email address, there's one extra step: you'll need to validate your address before you can finish creating your account. Look for an email in your inbox (and remember to check your spam folder if you don't see it). There's a code you can copy and paste into your browser window, or you can just click the link in the email to complete the validation process directly.

Once you've used an SSO system or validated your email address, there's one final important step: choosing a URL for your Grafana instance. Each instance of Grafana that is hosted in Grafana Cloud gets its own unique URL. This is what you'll put in your browser to go directly to your Grafana environment.

Grafana Cloud will suggest something for you based on your name. (If you have a common name, it might have to add a number or some extra text to make the URL unique.) But you can change this if you like, and the URL doesn't have to be based on your name at all. If you choose a URL that's already in use, Grafana Cloud will let you know that you need to try something else before you can continue. And note that there are a few restrictions on what you can use in the name. You can only use letters and numbers, no spaces, dashes, underscores, or periods. Figure 1-3 shows a valid URL being selected.

Figure 1-3. *Picking a URL*

Caution   Your URL can't be changed later, so be sure you're happy with it before you move on! Double-check your spelling to verify that it's 100% correct.

Once you click this last button, Grafana Cloud will create an instance of Grafana just for you. You'll be set up as the administrator for this instance, so you'll be able to configure data sources, set permissions on dashboards, and invite other users to the account. This may take a few seconds, but once it's finished, you'll be automatically logged in to your brand-new Grafana instance. When you see a getting started screen with a navigation bar on the left, you'll know your instance is running. When your instance starts, you'll see a screen like Figure 1-4.

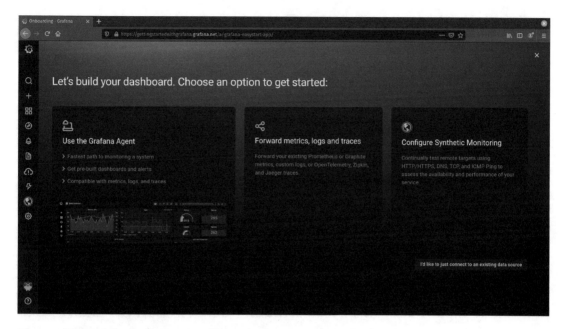

Figure 1-4. *A brand-new Grafana instance ready to use*

Congratulations! Your Grafana Cloud account and new Grafana instance are ready to use!

A Quick Overview of the Grafana Interface

We'll review the functionality of Grafana in more detail in the following sections. But it's worth taking a quick tour of the interface (Figure 1-5) to understand how the Grafana UI works and where to look for the tools you need.

Figure 1-5. *An example Grafana dashboard*

Navigating Grafana

On the left is a menu bar with a set of icons running top to bottom. This is the Grafana *navigation bar*, which provides quick links to the main functionality provided by Grafana. The stylized orange "G" at the top of the navigation bar is the Grafana logo, and clicking this will always take you to your Grafana home page. By default this page is an overview of the dashboards and alerts you've configured plus helpful links to install plugins and manage your Grafana Cloud account. But this can be customized or can be set to any dashboard you like.

The other buttons down the navigation bar will help you find or create dashboards, explore data, create and manage reports and alerts, manage data sources and other settings, and expose extra functionality provided by plugins.

At the very bottom of the navigation bar are two icons. The top is a user icon where you can manage your personal account settings. The bottom is a help menu that provides keyboard shortcuts and a way to request support from Grafana Labs.

Dashboards and Panels

Dashboards are the most important part of Grafana and as such take up most of the space in the interface. Everything outside of the dashboard controls at the top of the screen and the navigation bar is the dashboard space.

A dashboard is a collection of *panels*. Panels are the actual charts, graphs, numbers, tables, and other objects that show data. By design, Grafana can use any panel with any data source, so regardless of where your data is, you have the same options for visualizing it. This means that you can easily combine data from multiple different sources and still have a uniform representation of everything in one place.

Note   Panels can only show data that makes sense for the type of visualization they provide. If you have a panel that represents geographic location data, it probably won't show much if you give it quarterly sales figures represented in dollars.

There are many types of panels built into Grafana that will probably cover most of how you want to represent data. But you can also extend Grafana by adding panel plugins that provide different ways of visualizing your data. We'll look at more of this functionality in the next chapter.

Dashboard Options

Across the top of the screen whenever a dashboard is open (which is most of the time), you'll see a set of dashboard controls.

The top left shows the current dashboard folder and name. Grafana uses folders to organize dashboards for easy search and to let you group dashboards logically. Clicking the folder name will show you other dashboards in the same folder as the current dashboard. Clicking the dashboard name will show the full list of folders and a search field for quick dashboard access.

At the top right are two icons. The leftmost icon in the shape of a gear opens the *dashboard settings* menu. This includes metadata like the dashboard name and folder, variables that can be used to change functionality, how frequently the data on a dashboard should refresh, and other settings. We'll cover these in a future section.

The rightmost icon in the shape of a display changes Grafana's *view mode*. By default, the full Grafana UI is shown. But clicking this button lets you cycle through various display modes to hide the navigation bar or even put the dashboard into a full screen "kiosk" mode.

Creating Your First Dashboard

Now that you have a Grafana instance up and running, it's time to see what it can do! There are a few links on the default page that's created with your Grafana instance that will help you import data from various sources, but we'll start with some random sample data for now while you get a feel for using Grafana.

Adding Panels

Let's build a dashboard. Start by clicking the large + button on the navigation bar. This is the "new dashboard" button and will start creating a dashboard from scratch. You'll see that Grafana creates a new dashboard and starts by asking how you'd like to add your first panel, shown in Figure 1-6.

Figure 1-6. *Creating a new dashboard*

For now, click "Add an empty panel." This will create a new panel and bring up the panel editing interface (Figure 1-7).

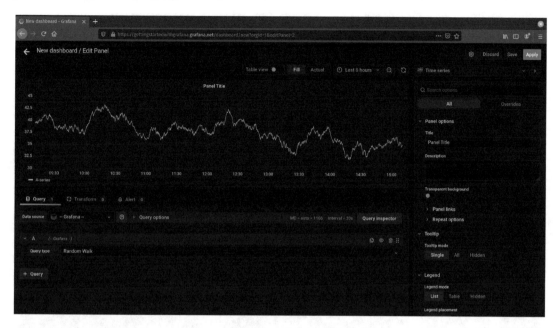

Figure 1-7. *A new panel with sample data*

There's a lot going on here! The panel editor provides a huge list of ways to query and manipulate data. We'll cover all of this in detail in the next chapter, so don't worry about all the options here yet. For now, just click the "Apply" button in the upper-right corner of the screen.

Once you apply the panel settings, you'll see that the panel has been placed on the dashboard. It's now showing the sample data in a simple line graph.

Let's add one more panel so we have a more interesting dashboard. Click the "Add panel" button in the upper right of the screen. (It looks like a bar graph with a large + next to it, as shown in Figure 1-8.) Once again, add an empty panel to the dashboard.

Figure 1-8. *The "Add panel" button*

This time, let's change the visualization from a time series to a gauge. Click the dropdown in the upper-right corner of the panel editor – it should currently say "Time series." Select "Gauge" and you'll immediately see the visual representation of the sample data change, as in Figure 1-9.

Figure 1-9. *Changing the visualization to a gauge*

Now hit "Apply" again, and you'll see the gauge added to your new dashboard.

Working with Panels

You can change the size and layout of panels on a dashboard just by clicking and dragging. To move a panel on the dashboard, click and hold on the *panel title* (the bar at the top of the panel) and drag the panel to where you want it to go. You'll see that other panels will move out of the way and rearrange themselves as you do this. This is because dashboards in Grafana are arranged in a grid. Panels will automatically snap to the closest grid location, ensuring that nothing gets lost or hidden behind another panel.

Resizing panels works similarly. Hover over the lower-right corner of a panel, and you'll see a small arrow appear. You can click and drag there to resize the panel. The panel visualization will resize to fit the new panel size. Some panels may also rearrange the data (or even show and hide components) as a panel grows or shrinks to make the best use of the available space.

Finally, if you want to change any of the options in your panel, just click the panel title. Select "Edit" from the menu that appears to go back into the panel options view (Figure 1-10).

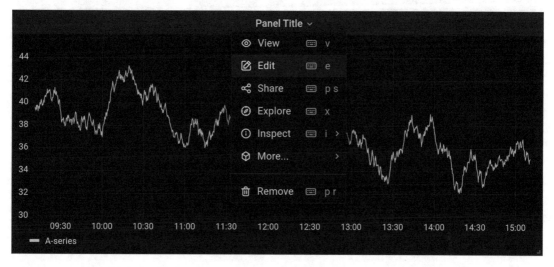

Figure 1-10. *The panel menu showing the "Edit" option highlighted*

Try creating a few panels and arranging them in different ways to see how your dashboard can be laid out. When you're happy with what you've built, click the "Save" icon at the top of the screen (Figure 1-11) to save your dashboard.

Figure 1-11. *The "Save" button*

Give your dashboard a descriptive name. You can sort the dashboard into a folder to keep it grouped with other related dashboards, but for now just keep the default "General" folder as shown in Figure 1-12. Click "Save" and your dashboard will stick around for next time!

Figure 1-12. *Saving a dashboard*

Logging Out of and In to Grafana

Creating an account on Grafana Cloud automatically logged you in to your new Grafana instance. In case you're on a shared computer or want to log out to switch accounts (or for any other reason), you can do that right from the Grafana window itself.

Signing Out

To log out, hover over your user icon in the lower left on the navigation bar. A menu will appear with your username and several options (Figure 1-13). Click "Sign out" and you'll be taken back to the Grafana login page for your instance.

Figure 1-13. *The "Sign out" option is available in your user menu*

Tip Logging out of your Grafana instance doesn't automatically log you out of Grafana.com. You can have access to more than one Grafana instance with a single Grafana.com account. The two accounts are connected but not the same. If it's a bit confusing, don't worry – you'll only need to think about any of this if you set up multiple Grafana instances from a single Grafana.com account. Otherwise, you can ignore the distinction and just think of your Grafana.com account as unlocking your Grafana instance whenever you log in.

Signing In

There are a couple of ways of signing back in to Grafana. The easiest is to simply visit your URL in a browser and click the button marked "Sign in with Grafana.com". If you're still logged in to your Grafana.com account, this will log you straight in to your Grafana instance as well. Otherwise, you'll see a login screen similar to the account creation screen from earlier like the one in Figure 1-14.

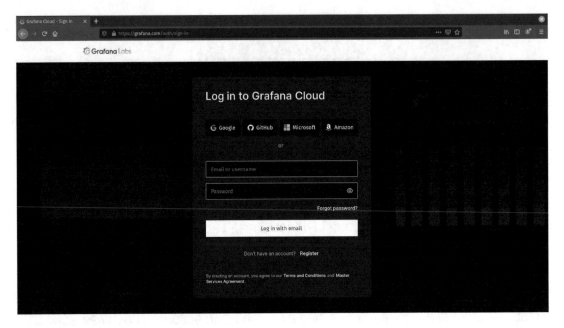

Figure 1-14. *The Grafana Cloud sign-in page*

Use the same option to log in here that you picked when you created your Grafana instance earlier, and you'll be brought straight to your Grafana home dashboard.

The other way to connect to your Grafana instance is through Grafana.com. Instead of going directly to your Grafana instance URL (e.g., *https://gettingstartedwithgrafana.grafana.net*), you can first log in to the Grafana.com management portal and access your Grafana instance there.

To access this portal, start by going to the Grafana home page at *www.grafana.com*. If you're not already logged in, click the "Login" link in the upper right and use the credentials for the account you created earlier. If you are already logged in, click the "My Account" link in the upper right. This will bring you to your account management portal, shown in Figure 1-15.

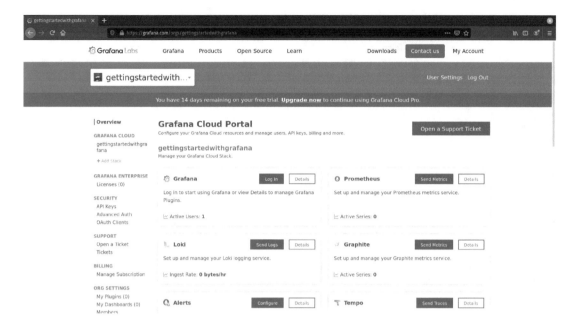

Figure 1-15. *The Grafana.com management portal*

There are a lot of options here, not all of which are directly relevant to Grafana. Feel free to explore and look at the other services and options available to you.

If you have access to multiple Grafana Cloud environments with one set of credentials, you can select from those environments using the dropdown menu at the top. If you've only created one, then you can ignore this menu.

To access your Grafana instance, look for the box labeled "Grafana" and click the "Log In" button.

Summary

In this chapter, you learned how to create a hosted Grafana instance on Grafana Cloud and used that instance to explore the main Grafana user interface. You created a dashboard and learned how to add panels to it, then how to arrange your dashboard by moving and resizing panels. Finally, you've seen how to manage your Grafana Cloud account and to log in to and out of Grafana.

In the next chapter, we'll cover working with panels in more depth, including various ways of representing and manipulating data.

## CHAPTER 2

Working with Panels

Grafana is built to be extensible, meaning that a wide variety of plugins exist to control how data is managed and visualized. You can even create your own plugins if you have needs that aren't met by what's available already. But that doesn't mean that it isn't useful without lots of extra effort! Grafana ships with a number of powerful visualizations natively, as well as functionality to let you run ad hoc queries to find the data you're looking for.

In this chapter, we'll take a look at the options available for controlling how data is displayed, some of the built-in visualizations available in Grafana, and how to find plugins to extend this functionality.

Tip When learning the capabilities of an expansive system like Grafana, it's usually best to do as much hands-on work as possible. That said, some of the examples here require specific types of data or extra setup to get good results. A great way to see what's possible with Grafana and the visualizations it provides is to visit *https://play.grafana.org/* where you'll find a large number of examples already built. You can click through these and even go into the panel edit view on dashboards to see how things work under the hood. It's a great resource, and I encourage you to look at it frequently for inspiration and help!

Basic Unit of Data Visualization

If you recall from Chapter 1, *panels* are the basic unit of data visualization in Grafana. Each panel shows data from at least one data source, though many panels can connect to more than one. How the panels display the data is up to the panel itself, but all panels provide a set of visualization options that give you as much control as possible.

Note   Grafana visualizes data but doesn't actually store it – it relies on external data sources to provide the actual data and a way to retrieve it. We'll learn more about working with data sources in Chapter 4. For now, just know that the sample data we're using for our panels is provided by a special built-in data source that generates random numbers.

We'll review this functionality with screenshots so you can just read straight through, but it's easy to follow along as well. The best way to do this is to start with a dashboard. You can either load up the dashboard you created in Chapter 1 or just click the large + on the Grafana navigation bar to create a new dashboard.

Once you have a dashboard open, click the **add panel** button in the upper right of the dashboard, and select "Add an empty panel" when the new panel appears. You can use this new panel to walk through the different visualization options and types.

Understanding the "Edit Panel" View

When you are editing a panel, Grafana uses the full browser window to show you only that panel. All other panels on the dashboard will be hidden while you focus on just this one.

The *edit panel* view is divided into three major parts: the visualization itself (and options for changing your view of it while editing), the query that is providing data for your visualization, and options for how to present this data on your dashboard. These components are shown in Figure 2-1 with each part outlined.

Figure 2-1. *The edit panel view*

The large part at the top left is the panel visualization itself. This shows the data that you are working with displayed by the panel type that you've selected. At the very top of this part are a few options to help you understand how your data will look when your dashboard is being used.

View Options

The first option, *table view*, lets you toggle between the visualization and the raw data in tabular format. This can be helpful if you have complex multidimensional data and you want to be certain that you're visualizing the right subset of it. You can flip back and forth between the visualization and the table view without having to run any extra queries or go directly to the data source itself to see the raw data.

Tip   The table view toggle only applies to the edit panel view. When you leave this view and go back to the dashboard, you'll see the actual visualization no matter if you had this option toggled on or off.

Next to the table view are two buttons labeled *fill* and *actual*. Because the edit panel view expands to fill the whole window, it can often be larger than the space allocated in the dashboard for the panel you're editing. By default, Grafana expands the panel to give you a larger view while you're working on it, but sometimes this isn't what you want. Some panels might change the flow or layout of data depending on the size, and others might have graphical items that expect an exact panel size. If you are concerned that your data will appear different on your dashboard than in this zoomed in view, click the "actual" button to make the visualization appear exactly the way it will on your dashboard. If you instead want to zoom in to experiment with additional detail, click the "fill" button to have the panel take all the available space.

Data Options

The next set of controls above the visualization itself control the amount of data that is displayed. Many common data sources use *time series* data – data that has a number of points over a length of time. Grafana provides options for changing the time range displayed on every dashboard. Since the viewers of your dashboard might change the time range they're looking at, it makes sense to be able to change that time range when editing a panel. This way, you can zoom in and out to make sure that your visualization makes sense no matter how long or short a period you're considering. Clicking this control opens a set of options to change this range, as shown in Figure 2-2.

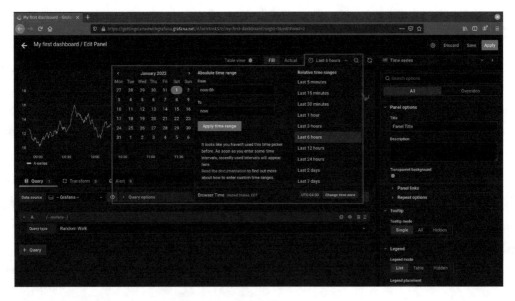

Figure 2-2. *The time range selection control*

You can select predefined time ranges like "last 5 minutes" or "last 7 days" from the menu, or you can specify exact times yourself. Grafana understands relative time (e.g., "now-24h" for the last day, "now-1h" for the past hour, etc.), or you can put exact dates (e.g., "2022-01-01 00:00:00" for midnight at the start of January 1, 2022).

Tip   If things look off, remember to check your time zone! It's often the case that data is stored in UTC but that you are viewing the data in another time zone entirely. Grafana tries to show you data in your local time zone, but if things are offset by several hours, it might be that the data is not in the time zone it expects. Click the "change time zone" button in the time range menu to move this.

Finally, to the right of the time range selector is a refresh button. Grafana will automatically refresh the data displayed when you change the time range, but if you're working with live data and want to see an event that happened since the last refresh, you can click this button to get the most up-to-date data.

Working with Data

Below the visualization itself are a set of tools for querying and modifying data. These tools will change depending on the data source, the structure of the data that the data source provides, and what actions you decide to take on that data.

The *query* tab controls how Grafana requests data from the data source. This view can change radically from one data source to another. Some may have a visual query builder, others may provide a set of dropdowns to filter data, and still others might simply have a text field for you to enter your query. Because each external tool has its own way of working with data, Grafana leaves this up to the data source itself. We'll look more at queries and data sources in Chapter 4.

The *transform* tab lets you manipulate the data once it's inside of Grafana. For example, you might want to change field names from a numeric ID to a more meaningful description, or reorder data, or even combine data from two sources to create a new metric. Transformations provide a powerful way to work with data inside of Grafana and will be covered in Chapter 9.

Finally, the *alert* tab controls alert conditions and responses. We'll explore alerting options in Chapter 11.

Panel Options

The right side of the edit panel view contains all the options available for your visualization. This includes the visualization type itself as a dropdown at the top and then all other options below that.

The panel options can be hidden by clicking the arrow on the far right of the selector, right next to the visualization name. This gives you more room to work with the query and see more data in the visualization itself. Clicking the arrow again will show the options.

You can also resize the options panel by using the mouse to drag the bar between the panel and the options pane. When customizing your panel options heavily, it's often useful to expand this panel so you can see more options at once.

Because each panel shows data in a different way, it doesn't make sense to try to force them all to have the exact same configuration options. For example, if you're showing time series data in a line chart, you want to be able to control things like whether lines are curved or sharp between points, whether the graph should be filled or a simple line, and whether to connect the lines around missing points of data. But if you're showing data as a table instead, there aren't any lines at all so none of those options make sense; instead, you'd want to control column or row headings, whether certain values should be colored differently, and how to format the values for currency or date formats. And all of that is out the window if you are showing geographic data on a map!

Because of the variety of ways of working with data, each visualization provides its own special options. We'll take a look at some of those options later in the chapter. But there are a set of options that are core to Grafana itself and provided for every panel type. Let's review those first.

Common Panel Options

The first section of options is titled simply "panel options" and controls how the panel is displayed on your dashboard.

The *title* is what your panel will be called. We've left this as the default of "Panel Title" so far, but you should set this for every panel you create. Make sure it's something descriptive enough that someone can tell what you are displaying without having to dive into the data itself. For example, "temperature per room" is a much better title than "degrees" as it provides more context about the data. (We'll explore this idea further in Chapter 6 when we talk about good dashboard design.)

Tip   Sometimes, you might not want a title at all, for example, if you're showing your company logo as a panel on a dashboard. In this case, it's fine to leave the title field blank. Grafana will shrink the panel title bar out of the way and give you a nice view for your graphic as shown in Figure 2-3.

Figure 2-3. *A panel without a title. Note the slightly larger visualization area*

The *description* field lets you give more information about what you're visualizing. This won't be displayed on the dashboard normally, but it will provide a tooltip that appears when you hover your mouse over the top left of the panel. This is useful for explaining more about the data or providing extra information that you don't want cluttering up the view at all times. Figure 2-4 shows the result of a description appearing when the mouse cursor hovers over the description icon on a panel.

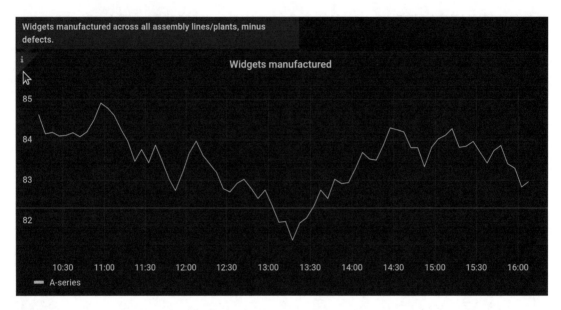

Figure 2-4. *A description displayed as a tooltip on a panel*

The *transparent background* toggle does just what it says. Normally, panels in Grafana have a solid background color (which might change depending on style settings or whether the viewer is using light mode or dark mode in Grafana). The panel will appear slightly lighter or darker than the rest of the dashboard so you can easily see the boundaries of the panel. Turning on the transparent background setting removes this background and places the panel directly onto the dashboard background. This is usually used for things like embedding logos onto a dashboard but can also help make data appear less or more important. Figure 2-5 shows a dashboard with both normal and transparent backgrounds. (This effect can be subtle. The default panels show a gray background in this example where the transparent panels show no background or outline at all.)

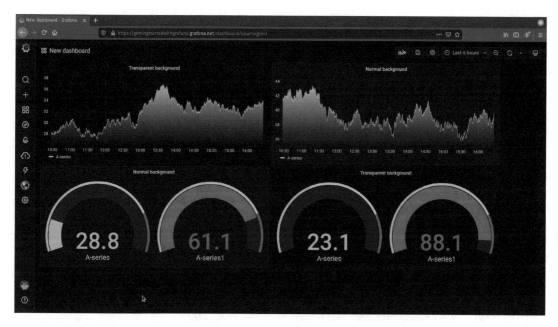

Figure 2-5. *Alternating normal and transparent panel backgrounds*

Panel links let you add clickable URLs to a panel. You can think of this as a set of bookmarks that are attached to a panel to give viewers of your dashboard quick access to other data or systems. These links can be to other Grafana dashboards, external systems, runbooks that provide troubleshooting instructions, or even automation systems to take some action straight from a dashboard – basically anything that your browser can access can be put here. Figure 2-6 shows a dashboard with two panel links, one to a more detailed dashboard and one to an external wiki. The links show up when hovering over the icon in the upper-left corner of the panel.

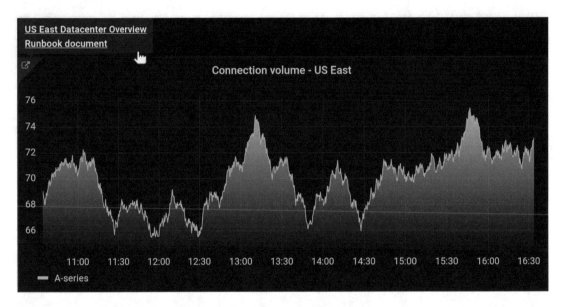

Figure 2-6. *Data links to internal and external resources*

Note   Panel links aren't the only way to add links to dashboards. You can add clickable links via the text panel which is described in the following section. Some panel types also provide functionality called *data links* which let you send additional data or parameters to your links. We'll look at panel links and data links in more detail in Chapter 7.

Finally, *repeat options* let you create a single panel but have it repeat multiple times for different values on your dashboard. This is advanced functionality that depends on dashboard variables, so we'll save this for Chapter 9.

Panel Types

There are too many ways to visualize data to possibly cover all of them in one place. This is especially true for a platform like Grafana where new visualizations are added frequently by the worldwide community of plugin developers. But there are some that you'll likely use over and over again that are built into Grafana directly. Let's review some of these panels and see what they can do.

> **Tip**   Some Grafana panels have a *lot* of options, way more than can be detailed here. Fortunately, they're all documented online at *https://grafana.* *com/docs/grafana/latest/visualizations/*. Be sure to check the documentation for more details about what you can tweak in each panel.

Time Series

The *time series* panel is the most frequently used visualization in most Grafana deployments, and so it is the default when you create a new panel. It gives you the ability to show lines, points, or bars with vertical height above or below the axis representing values and time shown horizontally from left to right. It's probably what you think about whenever someone says the word "graph" and is used for everything from stock prices to tracking daily weather data like temperature or rainfall. Figure 2-7 shows several possible representations of data all using the time series panel.

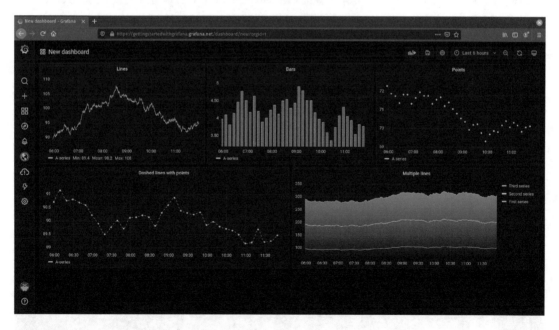

Figure 2-7.   *Multiple views of the time series panel*

The time series panel is useful when you want to show how a value is changing over time or to compare two or more values over time. It's great for showing data where the trend is the most important thing rather than one specific value. This is an easy one to play with as the sample data that Grafana generates works very well with this visualization, so be sure to play with the various options and experiment here!

Caution   While the time series panel gives you the ability to represent data as bars, this isn't really the same as a bar chart. Bars on a time series panel are still representing data over time with a bar for each point in time that data exists. To compare multiple groups of data without splitting values up over time, use another visualization like the *bar chart* or *pie chart* panels.

Bar Chart

The *bar chart* panel behaves very differently from the time series panel in that it represents data *categorically*. That means that instead of representing trends of data moving over a window of time, the bar chart panel compares groups of data points to one another in categories. For example, you might want to use the bar chart panel to compare votes for the best flavors of food – it's not data that changes over time, but you might see that chocolate is more popular for ice cream, while vanilla is more popular for cakes. Figure 2-8 shows this and several other data sets all represented with the bar chart visualization.

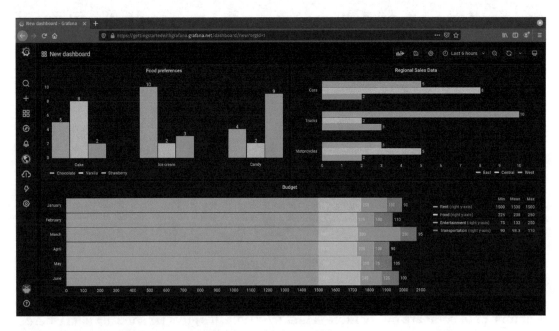

Figure 2-8. *Multiple views of the bar chart panel*

The default random data source won't work as well for bar chart data as it creates time series data. You'll need to add a data source that supports categorical data first, like a SQL database. The good news is that there's another more powerful data generator also built into Grafana called *Testdata DB* – it just needs to be enabled. We'll cover enabling this and connecting to other data sources in Chapter 4.

Stat

The *stat* panel is similar to the time series panel in that it shows time series data. In fact, by default it even shows a trend line over time just like the time series visualization. The key difference is that while the time series panel treats the trend as the most important information, the stat panel highlights a single value for display. By default, this is the most recent value but can also be the minimum, maximum, average, or another calculation. (You can even disable the trend line and show only the value itself in the *graph mode* options section of the panel.)

The stat visualization is most useful where you want to see what the most important value of a metric is right now. Things like the current number of logged in users of your web application or the number of manufacturing defects you've caught today are good candidates for the stat panel. It's intended to show data where the trend is interesting, the value is most important, but where you don't particularly care where the value sits in an expected range. The stat panel can also color the text or the background of the panel based on the value that is displayed as illustrated in Figure 2-9. Colors and their associated thresholds can be set in the stat panel options under the *thresholds* section.

Figure 2-9. *Two stat panels showing three time series*

Gauge and Bar Gauge

If you do care about ranges – and especially if a metric value is outside of that range – then the *gauge* or the *bar gauge* panel is a better choice. Like the stat panel, gauges show the current value of a metric. But they also let you specify an expected minimum and maximum value and show where within that range the value sits.

Gauges show their value as a pointer on a round dial, similar to a speedometer in a car. Bar gauges show their value as a bar extending from the minimum to the maximum value, more like a loudness meter on a stereo. Gauges and bar gauges have several display modes, as shown in Figure 2-10.

Figure 2-10. *A gauge panel and various bar gauges*

Gauges are a great choice for things like equipment temperature where you know that being too cold or too hot will cause a malfunction. You can show at a glance where the current temperature sits in that range and know when you should start to worry.

Table

Table panels are probably the simplest visualization possible, as they just show the data collected as raw values in rows and columns. It's useful when you just want to provide the raw data without plotting it graphically. But even though the data isn't plotted, you can still highlight rows, columns, or cells. For example, you might want to provide raw financial data alongside your trending and analysis, but highlight values that are higher or lower than expected. Using panel overrides, which we'll look at in Chapter 9, you can even embed bar gauges in table cells if you want to show a graphical representation of your data inside of the table. Figure 2-11 shows some examples of various table configurations.

Figure 2-11. *Table panels*

Tip   Table panels can only show two-dimensional data, so if you have more complex data you'll need to choose which series of data to show from a dropdown menu under the table when editing your panel. Grafana provides *transformations* to consolidate and reduce data that can help you show all the data you want to. We'll explore these in Chapter 9.

Pie Chart

Probably no other data visualization provokes stronger feelings than the *pie chart*. Data scientists publish endless blog posts about it being the worst way to show data, but it seems like no business presentation is complete without one. Love it or hate it, it's part of Grafana.

Pie charts can be used to show how a small number of items combine to make a whole and give you a rough idea of their proportionate size. Grafana offers you the option of a traditional pie chart or lets you cut out the center to make a donut-shaped version. Figure 2-12 illustrates both and shows some labeling options that can be set.

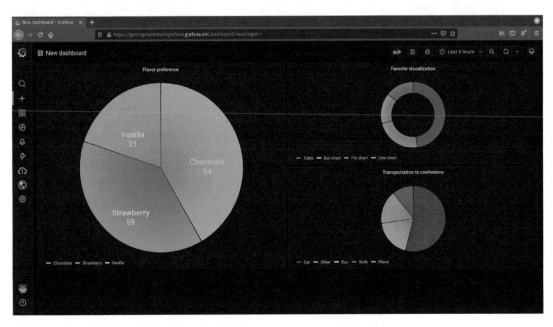

Figure 2-12. *Pie and donut charts*

State Timeline and Status History

The *state timeline* panel and the *status history* panel are closely related. Each one shows metrics on a timeline of values or states which can be defined by the data (like on/off, ok/warning/critical, occupied/empty, etc.). Where they differ is how the data is displayed.

The state timeline shows a continuous timeline of values, letting you see at a glance where a change occurred or how long the measurement stayed in that state. The status history shows a series of discrete measurements at regular time intervals and what the state of each metric was at that time.

Both can be used to show how the state of a set of metrics has changed over time as shown in Figure 2-13.

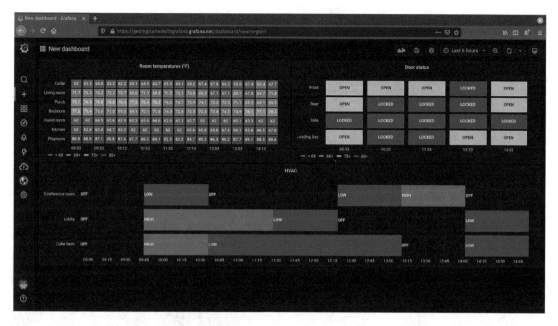

Figure 2-13. *State timeline and status history panels*

Heatmap and Histogram

Another set of closely related data visualizations, *histograms* show how data is distributed across the set of values received, while *heatmaps* show multiple histograms over time. Think of dividing your data into "buckets" that hold ranges of values like 0–10, 11–20, and so on. Histograms and heatmaps show how many data points are in each bucket.

A histogram shows the distribution of data. The famous "bell curve" shape that tends to arise in things like population studies and probability analysis is an example of a histogram where most of the data is in the middle buckets with comparatively fewer points at either edge. Histograms can show things like the most common amount of time taken for a web server to return a requested page.

Heatmaps take a series of histograms over time and represent them in a series. Because the horizontal axis is now used for time instead of buckets, the size of each bucket is represented with color or brightness. Generally, more values in a bucket are represented with brighter colors and/or moving from blue or green to red, though these are all configurable in the panel.

Figure 2-14 illustrates a histogram and two heatmaps.

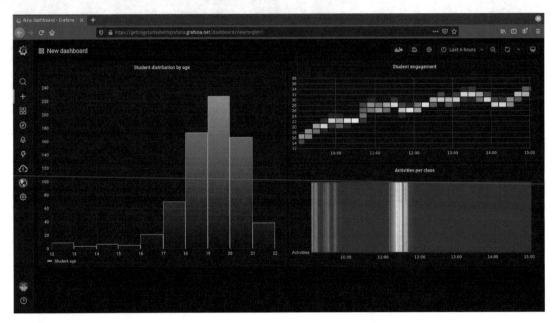

Figure 2-14. *Histogram and heatmap panels*

Text

At first glance, the *text* panel seems trivial – just a way to show plain text on a dashboard for notes or descriptions of panels. And while it certainly can do that, it's capable of much more.

By default, the text panel expects you to provide it with some *Markdown* formatted text. Markdown is a simple text formatting system used by many popular tools and web environments to make customizing the way text is displayed easier. It allows you to make lists, bold or italic text, change font size, and create links without writing full HTML. Grafana will convert your Markdown text to HTML transparently so you can make your dashboards beautiful. It even provides a link to some Markdown help at `https://commonmark.org/help/` if you want to learn more.

But Markdown isn't the only thing the text panel can understand. By changing a setting in the panel options, you can include full HTML in the text panel. This means that you can add images, use different fonts, add tables, change layouts – almost anything that you can do in a web page can be done inside of a text panel. This functionality makes it incredibly powerful when you want to quickly extend a Grafana dashboard beyond its existing capabilities without writing a whole new visualization plugin.

Caution   The text panel is capable of things like running JavaScript or embedding external web pages inside an IFrame, though this functionality is disabled by default due to security concerns. If you want to enable this in your Grafana environment, you'll need to turn on the *disable_sanitize_html* configuration option. More information about this is available in the Grafana documentation at *https://grafana.com/docs/grafana/latest/administration/ configuration/#disable_sanitize_html*.

Geomap

Most of the other panels we've looked at show data in relation to time or other data. The *geomap* panel, however, shows data in relation to a place.

Geomap visualizations show where (and how much) something has been measured on a map. By default, this is on a map of Earth, and the geomap panel provides several representations of Earth for you to work with. These include a simple global map, street data provided by the Open Street Map project, and cartographic and Geographic Information System (GIS) maps. But you can also provide your own base layer data, for example, if you wanted to show the perimeter of an area covered by sensors with specific and relevant markers already added.

The geomap panel allows you to zoom in and out and can add points to the graph based on a variety of methods including latitude/longitude pairs, geohash (a single value that represents a location on Earth), coordinates provided by your data source, or even an external lookup based on a given location name.

Because geographic data is often more complex than time series data, there is a lot that needs to be set up to get a good visualization. Reading the documentation is more important for this panel than any other. But the reward for this work is beautiful and flexible maps of your data that can update in real time as shown in Figure 2-15.

Figure 2-15. *The geomap panel showing heatmap data overlaid on a map*

Node Graph

The *node graph* panel is another way to represent data with a non-time dimension, in this case networks (or to use a more precise mathematical term, *directed graphs*).

The node graph shows individual values from objects in your data as well as how those objects are connected to one another. For example, you might have an assembly line that is building widgets. A given step in that assembly line might have inputs from previous steps (e.g., multiple components that are assembled to build this part of the widget) as well as an output that becomes an input for later steps.

The node graph allows you to represent the state of all the parts of your network and shows how each connects to the others. So if something near the beginning of your assembly line breaks, you can see everything that is impacted downstream as a result. Figure 2-16 shows a set of services and dependencies represented in a node graph.

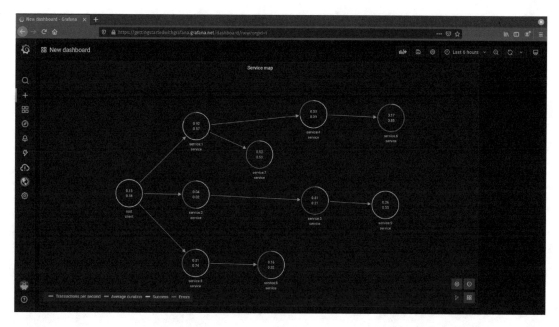

Figure 2-16. *A node graph panel showing service dependencies and status*

Others

The above is a sampling of the most useful and interesting visualizations shipped with Grafana, but is by no means everything. It's worth taking some time to browse the other panels that are included by default. There are built-in panels to show news or RSS feeds, active alerts that have triggered in your environment, non-graphable data like logs, even lists of related dashboards.

And that doesn't include all the additional visualizations that are available in the Grafana community! Let's take a look at how to find those next.

Finding Other Visualizations

Because visualizations in Grafana are plugins, you're not limited to just the options that ship by default. There are a variety of other ways of representing information on a dashboard, ranging from "traffic light" charts to show good/warning/critical status, interactive workflow diagrams that can navigate through complex systems while showing real-time measurements, and even animated maps that can display data over time and space like weather maps.

To see the available plugins, go to *www.grafana.com/plugins* and select "panel" as the type of plugins to view. This will provide a list of available panel plugins for you to install, as shown in Figure 2-17.

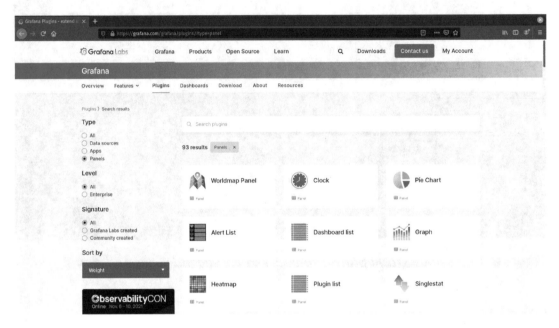

Figure 2-17. *The list of panel plugins available on the Grafana website*

Note   You can create your own visualizations and even publish them for others to use if you like! While explaining that is beyond the scope of this book, a good place to start is with the Grafana panel tutorial at *https://grafana.com/ tutorials/build-a-panel-plugin/*.

Summary

In this chapter, you learned how to manage panels on a Grafana dashboard. This includes options that are common to all panels as well as where to find options specific to each type of visualization. You have taken a tour through many of the visualizations that ship with Grafana and learned where each is useful and seen a few ways that they can represent data. You've also learned where to go to find more plugins to expand on these visualizations if you need to represent data in a way that isn't included in Grafana already.

In Part II (Chapters 3–5), we'll dive deeper into deploying and managing Grafana. We'll start by learning how to run Grafana outside of the cloud. From there, we'll see how to connect real data sources into Grafana to start visualizing your own data. At the end of Part II, we'll look at ways of managing users in a shared Grafana environment.

PART II

Deploying and Managing Grafana

Using a cloud service is the fastest way to get started, but not everyone wants to have their infrastructure managed by someone else. You might not want to rely on a third-party vendor, for example, or maybe you just want to understand how the system works in order to tinker with it for fun.

But this can also be driven by security or legal considerations. If your data sources aren't connected to the Internet, then a cloud-based solution won't be useful to you no matter how convenient it is. Having the option to run your own Grafana instance inside of a disconnected environment lets you visualize your data while still meeting your isolation requirements.

Regardless of where Grafana is deployed, you'll need to be able to manage things like connections to your data, user accounts and permissions, and possibly even external authentication services like Active Directory.

In Part II (Chapters 3–5), we'll look at how to deploy Grafana locally and how to administer it regardless of where it's running.

CHAPTER 3

Deploying Grafana Locally

Grafana is intentionally built to be both *self-contained* and *portable*. This means that it runs with few *external dependencies* (additional libraries or software packages that are not already part of most operating systems) and that it can run on a wide variety of hardware and operating system platforms.

This is great news if you're looking to run Grafana on your own, as this means that you don't need to buy special hardware or install a specific version of an operating system that you might not be familiar with. Grafana can run on your laptop just as happily as it will run in a large cloud environment. (It will even run on something as small as a Raspberry Pi!)

When you download Grafana, you'll see that it's provided in two editions, *OSS* and *Enterprise*. OSS refers to Open Source Software, meaning a completely open edition of Grafana with full source code available. The Grafana Enterprise package provides more functionality than the pure open source Grafana package, though it requires a license key for the extra features to be enabled. Without the license, the Grafana Enterprise package will function exactly the same as the open source package.

It's easiest to just install the Enterprise package when starting, as it makes unlocking these features in the future easier. However, you can always switch from one to the other without losing your data, so don't worry too much about which one you choose for now. We'll look at the features provided by Grafana Enterprise in Chapter 13.

Aside from picking which edition of Grafana you want, you won't need to spend a lot of time setting up support systems or other software to get going. Once you download and install the Grafana package, you'll be ready to go!

© Ronald McCollam 2022
R. McCollam, *Getting Started with Grafana*, https://doi.org/10.1007/978-1-4842-8309-7_3

Note   Grafana actually requires a SQL database to store things like settings, user account information, dashboard layouts, and other configuration data. One way that Grafana stays self-contained is by shipping with a lightweight file-based database called SQLite. SQLite is great for small environments, but when you start to think about things like high availability or larger scale for your Grafana environment, you'll want to use a more full-featured database system. We'll look at using an external database with Grafana in Chapter 11, but for now you don't need to worry about it – SQLite is more than powerful enough for a small Grafana deployment.

The installation and startup process for Grafana is slightly different for each platform. We'll take a look at some of the most common environments and walk through the setup process for each. And note that while you can run Grafana completely disconnected from the Internet, you will need to be connected to download the packages. If you're installing Grafana to a system that will never be connected to the Internet, you'll need to download the packages separately and copy them to your system yourself before proceeding.

Linux

Grafana is an open source tool that's widely used in IT monitoring, so it makes sense that the most common platform for Grafana is deployed on is an open source tool that's widely used in IT generally: Linux.

Caution   Many Linux distributions provide their own Grafana packages, so you may be able to install Grafana from a package store or software repository. However, these packages are not necessarily always up to date with the latest release of Grafana and may contain changes that make getting support more difficult. Additionally, if you're looking for some of the advanced features provided by Grafana Enterprise, those won't be provided by the packages provided by your Linux distribution. So it's always best practice to download Grafana directly from the source.

Linux comes in many flavors, each with its own idiosyncrasies that can change how installation works. We'll look at how to use installation packages for the two most popular "families" of Linux distributions, Debian and Red Hat, as well as a non-packaged method that should work for any Linux environment.

Debian (Including Ubuntu)

Debian is one of the two most common families of Linux. While Debian is often used itself, there are also a number of derivatives of Debian that use the same package format and configuration system, including Ubuntu.

To download Grafana, point your browser to *www.grafana.com* and click the "download" link at the top of the page. (You can also go directly to *www.grafana.com/ get*.) By default, you'll see information about using Grafana Cloud, so click the "self-managed" tab to see the links to download Grafana directly. Click the "Download Grafana" link as shown in Figure 3-1.

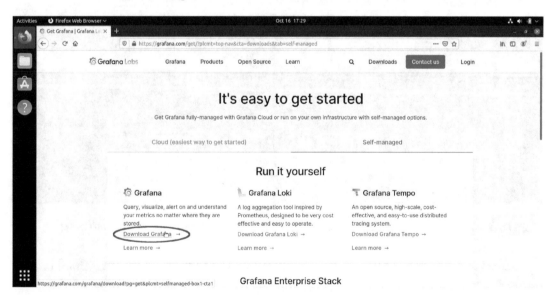

Figure 3-1. *Navigating to download Grafana from an Ubuntu Linux system*

After clicking the download link, you'll see the various download options available. Be sure the Linux version is selected (which should be the default), and the download links for the Linux versions of Grafana will be shown. By default, the latest released version of Grafana will be selected. If you need another version (such as an older release for compatibility, or you want to try a beta version), you can select it here.

Grafana Labs provides Debian packages for both pure open source Grafana and Grafana Enterprise, so you can select either edition to download here.

Once you've selected your Grafana version and edition, you'll see some commands to copy and paste lower in the page. To run these, they'll need to be pasted into a terminal.

If you're connected to a remote Linux system through *ssh* or *telnet* (or some other type of remote connection that uses a command line), you can paste those commands here. If you're using a Linux desktop, you'll first need to open a terminal window. How you open a terminal window will vary somewhat depending on your Linux environment, but usually you will find this in a system menu. Figure 3-2 shows where to find the terminal launcher after clicking the menu button in the Ubuntu desktop.

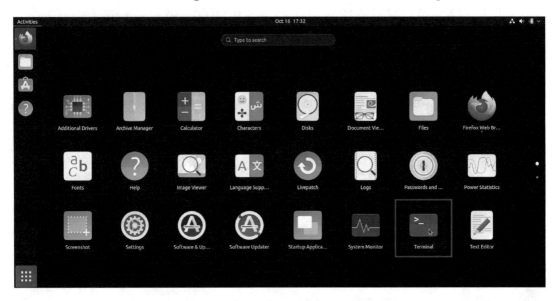

Figure 3-2. *The Ubuntu application menu with the terminal icon outlined*

Once your terminal is open, copy the commands from the Grafana download page under the "Ubuntu and Debian" section and paste them into your terminal window. Listing 3-1 shows the commands to download the Debian version of Grafana Enterprise 8.2.1. Note that the command you will use may be different if you have selected a different version or edition of Grafana.

Listing 3-1. Installing Grafana in a Debian environment

```
sudo apt-get install -y adduser libfontconfig1 wget
wget https://dl.grafana.com/enterprise/release/grafana-enterprise_8.2.1_
amd64.deb
sudo dpkg -i grafana-enterprise_8.2.1_amd64.deb
```

Note   You may be prompted for a password after the first command, due to it running sudo to temporarily get escalated permissions. Just enter the password that you use to log in to the system here. If you don't have the appropriate permissions to install software, talk to your system administrator before continuing.

Let's step through these commands to understand what's happening.

While Grafana tries to limit dependencies, there are still a few things that it needs to function correctly that might not be installed on all systems. The first command installs the adduser and libfontconfig1 packages:

```
sudo apt-get install -y adduser libfontconfig1
```

These allow the Grafana package to set up a service account for Grafana to run as for security purposes and to install and configure fonts for rendering data. They may already be installed on your system, in which case this will just verify that and make no other changes.

Next is the actual download of Grafana:

```
wget https://dl.grafana.com/enterprise/release/grafana-enterprise_8.2.1_
amd64.deb
```

This command fetches the installation package for the Enterprise edition of Grafana 8.2.1.

Finally, we install the package using the Debian package manager, dpkg:

```
sudo dpkg -i grafana-enterprise_8.2.1_amd64.deb
```

You should see some informational output after each of these commands as shown in Figure 3-3. If there are errors, check your Internet connection and carefully read the error messages to troubleshoot the issues.

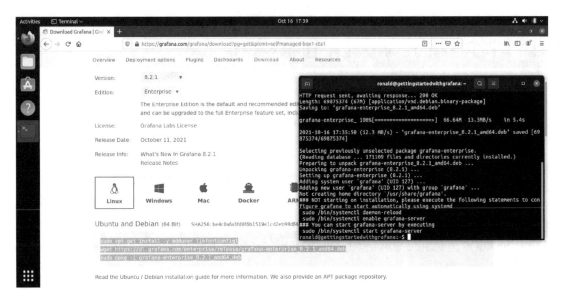

Figure 3-3. *A successful installation*

Once you've run the preceding commands, Grafana is installed! But it still isn't up and running yet. By default, Grafana does not start up automatically, as it's assumed that you might want to make some changes to its configuration to better suit your environment before using it. So we need to tell the Linux system to treat Grafana as a system service and to start it up so that we can use it.

Fortunately, the Grafana package tells us how to do this after installation. Listing 3-2 shows the relevant commands that the installer provides.

Listing 3-2. Enabling Grafana as a service and starting it up

```
sudo /bin/systemctl daemon-reload
sudo /bin/systemctl enable grafana-server
sudo /bin/systemctl start grafana-server
```

These commands work with systemd, a commonly used service management system for Linux. Grafana installs itself as a systemd service, so the first command tells systemd to reload its list of services so that it sees Grafana. Once systemd knows that the Grafana service is installed, we can tell systemd to enable it, which means to set it to start up whenever the system is booted. Of course, we don't really want to have to reboot the system just to start up a new service. So the last command is used to start the Grafana service and monitor it to be sure that it will stay up and running.

Now that Grafana is installed and started as a service, you can open a web browser and log in, as shown in Figure 3-4. The address you will use is *http://localhost:3000*. When Grafana is first started, there's a single administrator user created. The default username and password for this user are both *admin* (all lowercase).

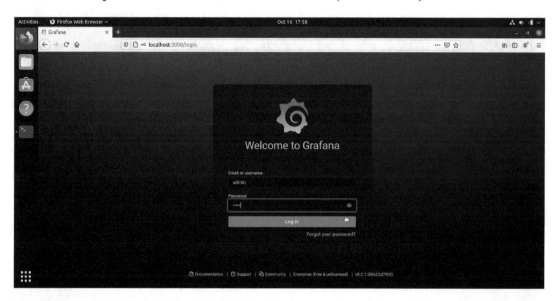

Figure 3-4. *Logging in to Grafana running on Ubuntu*

The first time you log in with the administrator account, you'll be prompted to change your password. Be sure to choose a secure password for this account as it has complete control of your Grafana environment.

Once you've changed your password, you're ready to start using Grafana! You'll probably want to skip ahead to Chapter 4 to start connecting data sources to make use of your new Grafana installation.

Red Hat (Including Fedora and CentOS)

Red Hat is a Linux distribution commonly used in larger enterprise companies and, like Debian, is the base for other distributions like Fedora and CentOS. All of these distributions use the same underlying RPM package format and use the same commands to install and run Grafana.

To download Grafana, point your browser to *www.grafana.com* and click the "download" link at the top of the page. (You can also go directly to *www.grafana.com/ get*.) By default, you'll see information about using Grafana Cloud, so click the "self-managed" tab to see the links to download Grafana directly. Figure 3-5 shows the download page on a Red Hat desktop, highlighting the "Download Grafana" link.

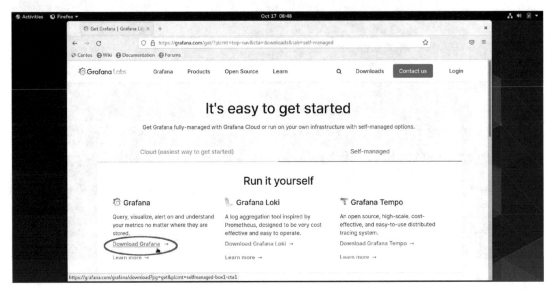

Figure 3-5. *The Grafana download page on a Red Hat desktop*

After clicking the download link, you'll see the various download options available. Be sure the Linux version is selected (which should be the default), and the download links for the Linux versions of Grafana will be shown. By default, the latest released version of Grafana will be selected. If you need another version (such as an older release for compatibility, or you want to try a beta version), you can select it here.

Grafana Labs provides Red Hat packages for both pure open source Grafana and Grafana Enterprise, so you can select either edition to download here.

Once you've selected your Grafana version and edition, you'll see some commands to copy and paste lower in the download page. To run these, they'll need to be pasted into a terminal. Be sure to copy the commands for Red Hat (rather than another Linux type) as shown in Figure 3-6. You may need to scroll down to find these commands.

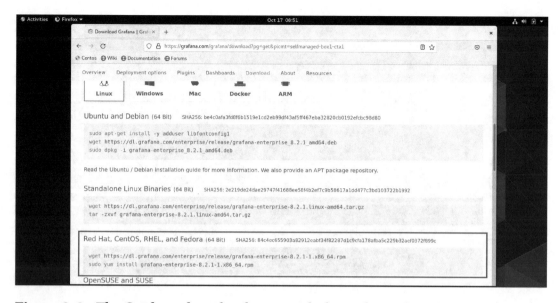

Figure 3-6. *The Grafana download page with the Red Hat download commands highlighted*

If you're connected to a remote Linux system through *ssh* or *telnet* (or some other type of remote connection that uses a command line), you can paste those commands here. If you're using a Linux desktop, you'll first need to open a terminal window. How you open a terminal window will vary somewhat depending on your Linux environment, but usually you will find this in a system menu. Figure 3-7 shows where to find the terminal launcher after clicking the "Activities" link at the upper right of the Red Hat desktop. Depending on how many programs you have installed, you may need to search for the terminal application to see it in the list.

Figure 3-7. *Opening the terminal window on a Red Hat desktop*

Once your terminal is open, copy the commands from the Grafana download page under the "Red Hat, CentOS, RHEL, and Fedora" section and paste them into the terminal window. Listing 3-3 shows the commands to download the Red Hat version of Grafana Enterprise 8.2.1. Note that the command you will use may be different if you have selected a different version or edition of Grafana.

Listing 3-3. Installing Grafana in a Red Hat environment

```
wget https://dl.grafana.com/enterprise/release/grafana-enterprise-8.2.1-1.
x86_64.rpm
sudo yum install grafana-enterprise-8.2.1-1.x86_64.rpm
```

Note   You may be prompted for a password after the second command, due to it running `sudo` to temporarily get escalated permissions. Just enter the password that you use to log in to the system here. If you don't have the appropriate permissions to install software, talk to your system administrator before continuing. You'll also need to have the `wget` program installed to download the Grafana package.

The first command, wget, downloads the Grafana package. The second command, yum install, installs Grafana in your Linux system.

As Figure 3-8 shows, you may be prompted to confirm the installation. Pressing "Y" will confirm that you do want to install Grafana. As the installation runs, you should see some informational output. If there are errors, check your Internet connection and carefully read the error messages to troubleshoot the issues.

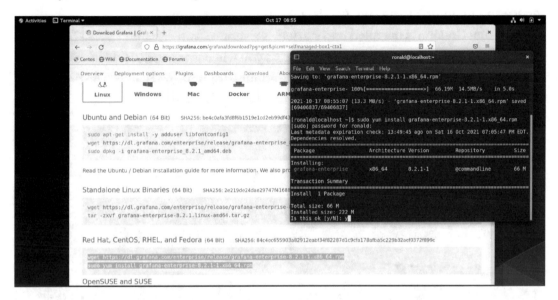

Figure 3-8. *Yum prompting to verify the installation of Grafana*

Once you've run the preceding commands, Grafana is installed! But it still isn't up and running yet. By default, Grafana does not start up automatically, as it's assumed that you might want to make some changes to its configuration to better suit your environment before using it. So we need to tell the Linux system to treat Grafana as a system service and to start it up so that we can use it.

Fortunately, the Grafana package tells us how to do this after installation. Listing 3-4 shows the relevant commands that the installer provides.

Listing 3-4. Enabling Grafana as a service and starting it up

```
sudo /bin/systemctl daemon-reload
sudo /bin/systemctl enable grafana-server.service
sudo /bin/systemctl start grafana-server.service
```

These commands work with `systemd`, a commonly used service management system for Linux. Grafana installs itself as a `systemd` service, so the first command tells `systemd` to reload its list of services so that it sees Grafana. Once `systemd` knows that the Grafana service is installed, we can tell `systemd` to `enable` it, which means to set it to start up whenever the system is booted. Of course, we don't really want to have to reboot the system just to start up a new service. So the last command is used to `start` the Grafana service and monitor it to be sure that it will stay up and running.

Now that Grafana is installed and started as a service, you can open a web browser and log in, as shown in Figure 3-9. The address you will use is *http://localhost:3000*. When Grafana is first started, there's a single administrator user created. The default username and password for this user are both *admin* (all lowercase).

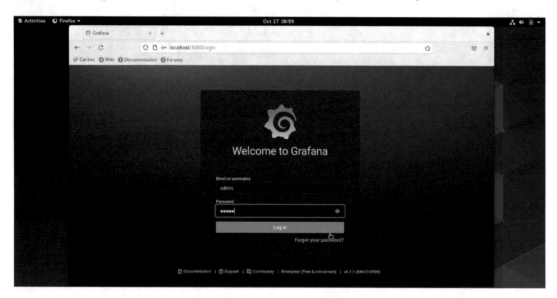

Figure 3-9. *Logging in to Grafana running on Red Hat*

The first time you log in with the administrator account, you'll be prompted to change your password. Be sure to choose a secure password for this account as it has complete control of your Grafana environment.

Once you've changed your password, you're ready to start using Grafana! You'll probably want to skip ahead to Chapter 4 to start connecting data sources to make use of your new Grafana installation.

Other Linux

While Debian and Red Hat are the most popular Linux distributions, they're not the only options out there. If you need to run Grafana in a different environment or just want to customize the installation to meet your needs, you can download the non-packaged binary version.

To find this version, first point your browser to *www.grafana.com* and click the "download" link at the top of the page. (You can also go directly to *www.grafana.com/get.*) By default, you'll see information about using Grafana Cloud, so click the "self-managed" tab to see the links to download Grafana directly. Figure 3-10 shows the download page, highlighting the "Download Grafana" link.

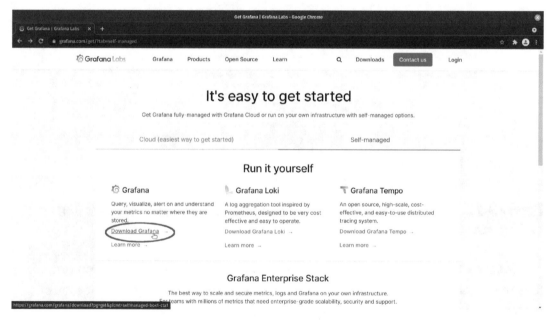

Figure 3-10. *The "Download Grafana" link*

After clicking the download link, you'll see the various download options available. Be sure the Linux version is selected (which should be the default), and the download links for the Linux versions of Grafana will be shown. By default, the latest released version of Grafana will be selected. If you need another version (such as an older release for compatibility, or you want to try a beta version), you can select it here.

Grafana Labs provides non-packaged archives for both pure open source Grafana and Grafana Enterprise, so you can select either edition to download here.

Once you've selected your Grafana version and edition, you'll see some commands to copy and paste lower in the download page. To run these, they'll need to be pasted into a terminal. Be sure to copy the commands in the "Standalone Linux Binaries" section (rather than another Linux type) as shown in Figure 3-11. You may need to scroll down to find these commands.

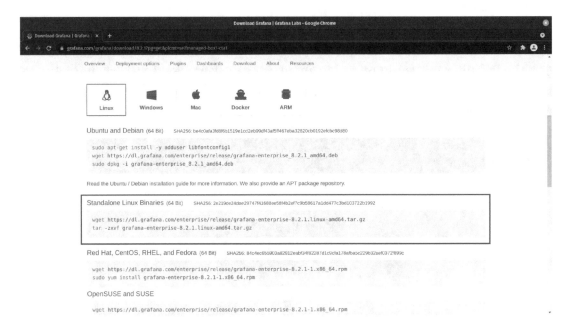

Figure 3-11. *The commands to download the non-packaged Linux version of Grafana*

Be sure you have the proper permission to write to the location where you run these commands. Your home folder is a good option, but depending on your Linux environment, you may want to deploy Grafana to another location.

Once you have copied the commands from the Grafana download page, paste them into a terminal on your Linux machine. Listing 3-5 shows the commands to download the unpackaged Linux version of Grafana Enterprise 8.2.1. Note that the command you will use may be different if you have selected a different version or edition of Grafana.

Listing 3-5. Downloading and uncompressing Grafana

```
wget https://dl.grafana.com/enterprise/release/grafana-
enterprise-8.2.1.linux-amd64.tar.gz
tar -zxvf grafana-enterprise-8.2.1.linux-amd64.tar.gz
```

The first command, wget, will download the archive containing the unpackaged version of Grafana. The second command, tar, will uncompress Grafana into its own folder.

At this point, Grafana is available on your system but is not yet running. Because it's a standalone binary rather than a package, you'll need to either configure it as a service in your Linux environment or start it manually. (Creating a service varies depending on your Linux environment, so consult your distribution's documentation for more information on this.)

To start Grafana manually, you just need to switch your current location to the Grafana folder and run the binary, as shown in Listing 3-6. Note that you will need to replace "8.2.1" with the version of Grafana you have downloaded.

Listing 3-6. Running standalone Grafana 8.2.1

```
cd grafana-8.2.1
./bin/grafana-server
```

The first command, cd, stands for "change directory" and will point your command line to your Grafana deployment. The second, ./bin/grafana-server, will actually run Grafana. Note the dot and slash (./) at the beginning of this command. If you leave them off it won't work, so be sure you include them!

Tip   If you forget the version number you downloaded, try running the ls command in the terminal. This will show the files and folders in your home folder, including your Grafana folder with the version number at the end.

As soon as you start up the Grafana server, you'll see a number of informational messages in your terminal as shown in Figure 3-12. Wait a few moments for these to stop scrolling. That will be your indication that Grafana has finished starting up and is ready to use.

Figure 3-12. *Output shown when Grafana is starting*

Once those messages stop scrolling, open up a web browser and point it to *http://localhost:3000*. When Grafana is first started, there's a single administrator user created. The default username and password for this user are both *admin* (all lowercase).

The first time you log in with the administrator account, you'll be prompted to change your password. Be sure to choose a secure password for this account as it has complete control of your Grafana environment.

Once you've changed your password, you're ready to start using Grafana! You'll probably want to skip ahead to Chapter 4 to start connecting data sources to make use of your new Grafana installation.

Caution   If you close the terminal window or connection you used to start Grafana, you'll cause Grafana to stop running as well. Be sure to leave this window open while you are using Grafana. If you accidentally close the terminal window, don't worry; you can just open a new one and rerun the two commands from Listing 3-6 to start it again.

Windows

While Linux is the most common platform for running Grafana servers, it's not the only supported environment. Windows desktops are commonly used for development, and Windows servers are often used for IT infrastructure, particularly in environments with large deployments of other Microsoft tools.

Grafana Labs provides both an installation package for Windows and a standalone binary. Most people will want the installation package as this will automatically install Grafana to the proper location and set up Grafana as a Windows service so that it can be managed easily. We'll focus on the installation package here.

To download Grafana, point your browser to *www.grafana.com* and click the "download" link at the top of the page. (You can also go directly to *www.grafana.com/get*.) By default, you'll see information about using Grafana Cloud, so click the "self-managed" tab to see the links to download Grafana directly. Figure 3-13 shows the download page on a Windows desktop, highlighting the "Download Grafana" link.

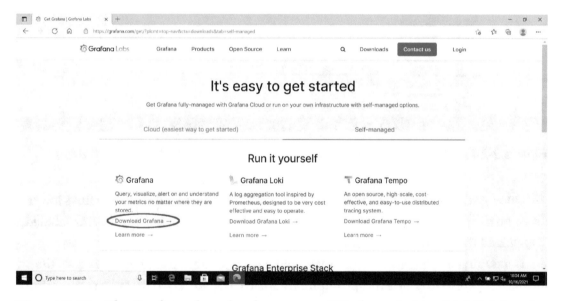

Figure 3-13. *The Grafana download page on a Windows desktop highlighting the download link*

After clicking the download link, you'll see the various download options available. Be sure the Windows version is selected by clicking the "Windows" button as shown in Figure 3-14. By default, the latest released version of Grafana will be selected. If you need another version (such as an older release for compatibility, or you want to try a beta version), you can select it here.

Grafana Labs provides Windows installers for both pure open source Grafana and Grafana Enterprise, so you can select either edition to download here.

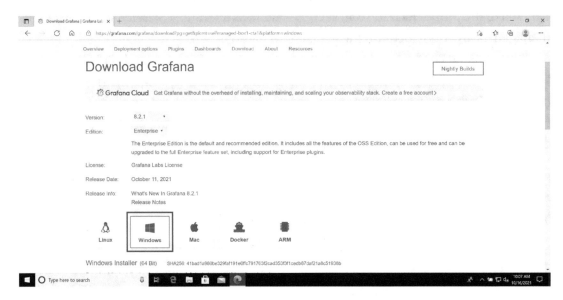

Figure 3-14. *The Grafana download page highlighting the Windows button*

Once you've selected your Grafana version and edition, you'll see some links lower down on the page. (You may need to scroll down to see them.) Look for the link labeled "Download the installer" and click it.

Depending on your security settings, you may be prompted to keep or discard the installer. Figure 3-15 shows this prompt. Be sure to click "Keep" so that you can continue the installation.

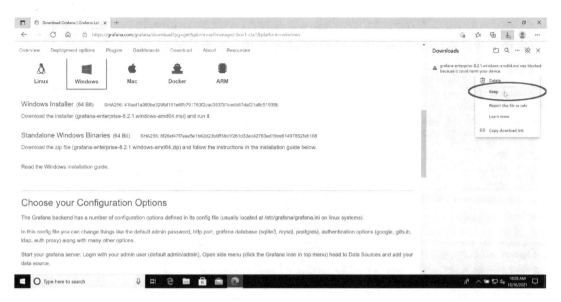

Figure 3-15. *The Grafana installer download warning with the "Keep" option highlighted*

After the download completes, select the "Open file" option under the "Downloads" menu for the Grafana installer. At this point, you may be prompted for security reasons again, as shown in Figure 3-16. If this prompt appears, start by clicking the "More info" link.

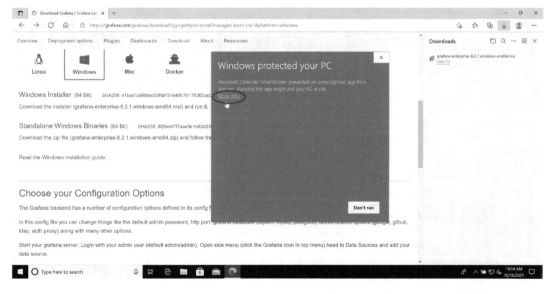

Figure 3-16. *The Windows security prompt highlighting the "More info" link*

Once you click "More info," you'll be presented with an option to run the installer. Click the "Run anyway" button at the bottom of the window as shown in Figure 3-17.

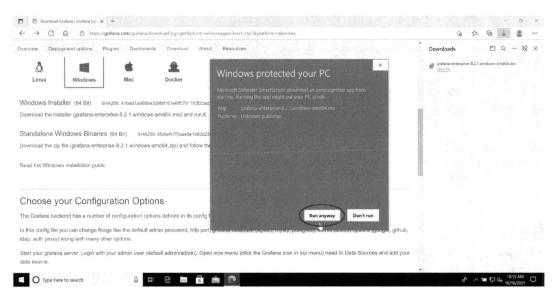

Figure 3-17. *The "Run anyway" button*

Finally, after navigating these security settings, the Grafana installation wizard will appear, which should look similar to Figure 3-18. Walk through this wizard as you would any other Windows installer by clicking "Next."

Figure 3-18. *The first page of the Grafana installation wizard*

There's one major choice to be aware of in this installation, which is whether you want to run Grafana as a Windows service. By default, this is enabled, as shown in Figure 3-19. This allows Grafana to start automatically when your system boots and enables you to control Grafana using the Windows service management system, so it's definitely recommended to leave this enabled. But if you prefer to run Grafana manually yourself, you can disable it here.

Figure 3-19. *The option to enable or disable Windows service integration for Grafana*

Once you complete the configuration and start the actual installation, you may be presented with yet one more security prompt, as shown in Figure 3-20. Click "Yes" here to allow Grafana to install itself to your computer.

Figure 3-20. *The final security prompt with the "Yes" button highlighted*

The Grafana installation should complete normally after this step. If you see any errors, run the installer again and be careful to select the proper security answers as outlined earlier.

Grafana is now installed! Assuming that you left the Windows service integration option turned on during installation, you can manage Grafana from the Windows "Services" control panel, as shown in Figure 3-21. Double-clicking the Grafana service will open a window where you can control whether it starts automatically and allows you to start and stop Grafana manually.

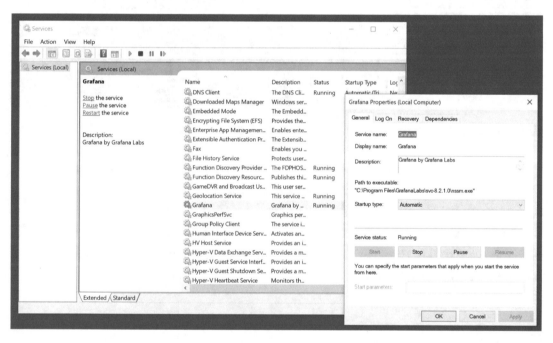

Figure 3-21. *The Windows "Services" control panel and the Grafana service control window*

To log in to your Grafana instance, open a web browser and go to *http://localhost:3000* as shown in Figure 3-22. When Grafana is first started, there's a single administrator user created. The default username and password for this user are both *admin* (all lowercase).

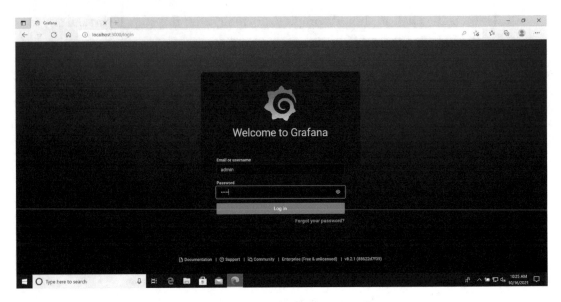

Figure 3-22. *Logging in to Grafana on a Windows desktop*

The first time you log in with the administrator account, you'll be prompted to change your password. Be sure to choose a secure password for this account as it has complete control of your Grafana environment.

Once you've changed your password, you're ready to start using Grafana! You'll probably want to skip ahead to Chapter 4 to start connecting data sources to make use of your new Grafana installation.

MacOS

While Linux and Windows are frequently used as server operating systems, there is no officially supported server version of MacOS. This means that it's rare to find Grafana running full time on a MacOS system serving up dashboards to multiple people. But if you run MacOS on your desktop or laptop, it can still be quite handy to run Grafana locally when you are testing or developing your observability and monitoring tools.

Fortunately, Grafana Labs provides a standalone version of Grafana for MacOS for exactly this reason. It's distributed as a single binary rather than as a full MacOS application, so if you're not used to using the command line in MacOS, it might take some getting used to. But it's not as hard as it sounds, so let's jump in!

Start by pointing your browser to *www.grafana.com* and then click the "download" link at the top of the page. (You can also go directly to *www.grafana.com/get*.) By default, you'll see information about using Grafana Cloud, so click the "self-managed" tab to see the links to download Grafana directly. Figure 3-23 shows the download page on a MacOS desktop, highlighting the "Download Grafana" link.

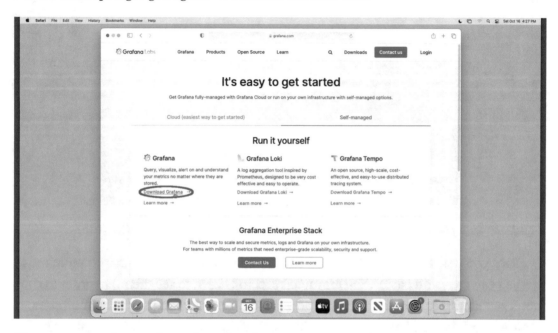

Figure 3-23. *The Grafana download page on a MacOS desktop highlighting the download link*

After clicking the download link, you'll see the various download options available. Be sure the MacOS version is selected by clicking the "Mac" button as shown in Figure 3-24. By default, the latest released version of Grafana will be selected. If you need another version (such as an older release for compatibility, or you want to try a beta version), you can select it here.

Grafana Labs provides an image for both pure open source Grafana and Grafana Enterprise, so you can select either edition to download here.

Figure 3-24. *The Grafana download page highlighting the Mac button*

Once you've selected your Grafana version and edition, you'll see some commands that you can copy and paste. (You may need to scroll down to see them.) In order to run these, you'll need to open the terminal application. You can find this by opening your Applications folder and navigating to Utilities. Double-click the application labeled "Terminal" as shown in Figure 3-25.

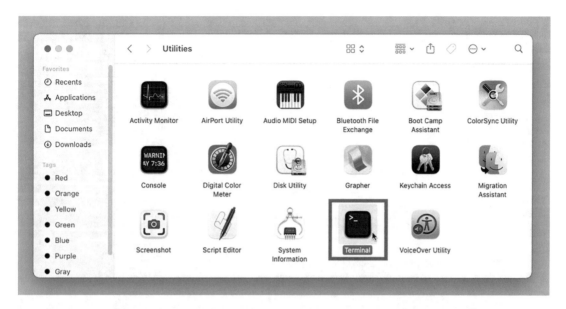

Figure 3-25. *The Utilities folder with the terminal application highlighted*

Once your terminal is open, copy the commands from the Grafana download page under the "Standalone MacOS/Darwin Binaries" section and paste them into the terminal window. Listing 3-7 shows the commands to download the MacOS version of Grafana Enterprise 8.2.1. Note that the command you will use may be different if you have selected a different version or edition of Grafana.

Listing 3-7. Installing Grafana in a MacOS environment

```
curl -O https://dl.grafana.com/enterprise/release/grafana-
enterprise-8.2.1.darwin-amd64.tar.gz
tar -zxvf grafana-enterprise-8.2.1.darwin-amd64.tar.gz
```

The first command, `curl`, downloads the Grafana package. The second command, `tar`, uncompresses the Grafana package into your home folder. Figure 3-26 shows the output that is produced by this command. It's normal to see a long string of filenames here. If instead you see an error, double-check your Internet connection and try copying the commands again.

Figure 3-26. *The output from extracting the Grafana package*

At this point, you have Grafana downloaded, but it's not yet running. To start it up now (and in the future if you stop Grafana or reboot), you'll need to run two more commands. Listing 3-8 shows the commands for the same 8.2.1 version of Grafana 8, but you might need to change the first command to match the version of Grafana you downloaded.

Listing 3-8. Running Grafana in MacOS

```
cd grafana-8.2.1
./bin/grafana-server
```

The first command, `cd`, stands for "change directory" and will point your command line to your Grafana deployment. The second, `./bin/grafana-server`, will actually run Grafana. Note the dot and slash (`./`) at the beginning of this command. If you leave them off, it won't work, so be sure you include them!

Tip   If you forget the version number you downloaded, try running the `ls` command in the terminal. This will show the files and folders in your home folder, including your Grafana folder with the version number at the end.

As soon as you start up the Grafana server, you'll see a number of informational messages in your terminal, as shown in Figure 3-27. Wait a few moments for these to stop scrolling. That will be your indication that Grafana has finished starting up and is ready to use.

Figure 3-27. *Informational messages produced when Grafana is starting*

Once those messages stop scrolling, open up a web browser and point it to *http://localhost:3000* as shown in Figure 3-28. When Grafana is first started, there's a single administrator user created. The default username and password for this user are both *admin* (all lowercase).

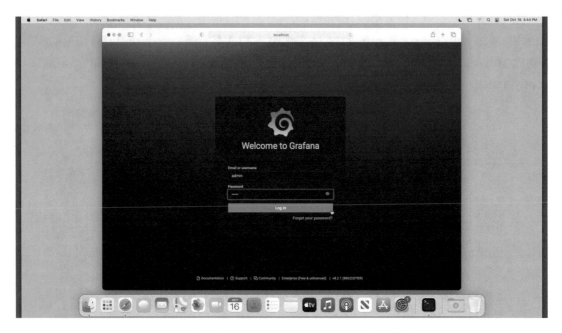

Figure 3-28. *Logging in to Grafana running on a MacOS desktop*

The first time you log in with the administrator account, you'll be prompted to change your password. Be sure to choose a secure password for this account as it has complete control of your Grafana environment.

Once you've changed your password, you're ready to start using Grafana! You'll probably want to skip ahead to Chapter 4 to start connecting data sources to make use of your new Grafana installation.

Caution   If you close the terminal window you used to start Grafana, you'll cause Grafana to stop running as well. Be sure to leave this window open while you are using Grafana. If you accidentally close the terminal window, don't worry; you can just open a new one and rerun the two commands from Listing 3-8 to start it again.

Docker

Docker is a popular system for building, running, and managing *containers*. A container is similar to a virtual machine in that it's self-contained and isolated from other environments running on the same physical system, but different in that containers are typically much smaller and lighter to run as they don't run a full operating system. Instead, they rely on the host OS providing most of the underlying system resources and libraries, so the container only needs to have the specific components that the application or service needs to run.

Using a Docker container is a bit different from other methods of installing Grafana. For one, it requires that you have the Docker engine set up and running on your local system. Docker is available for Linux, Windows, and MacOS, so you can run it on any platform you choose.

Another difference is that the Grafana Docker image is created by Grafana Labs but distributed through a central repository called Docker Hub. This means that you can easily install the Grafana Docker image without having to download it directly from the Grafana website.

Instructions for installing the Docker version of Grafana are available on the Grafana website. To find them, open a browser and go to *www.grafana.com* and click the "download" link at the top of the page. (You can also go directly to *www.grafana. com/get*.) By default, you'll see information about using Grafana Cloud, so click the "self-managed" tab to see the links to download Grafana directly. Click the "Download Grafana" link as shown in Figure 3-29.

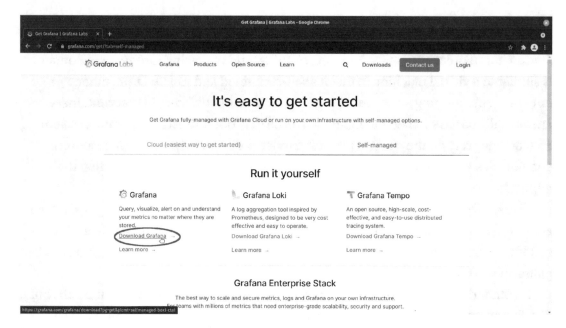

Figure 3-29. *The Grafana download page with the Download Grafana link highlighted*

Once you've clicked the download link, select the "Docker" button to show the Docker installation instructions, as shown in Figure 3-30. By default, the latest released version of Grafana will be selected. If you need another version (such as an older release for compatibility, or you want to try a beta version), you can select it here.

Grafana Labs provides an image for both pure open source Grafana and Grafana Enterprise, so you can select either edition to download here. You'll also have the choice of which base image is used inside the Docker container. Either image will work, though the Alpine image is smaller to download.

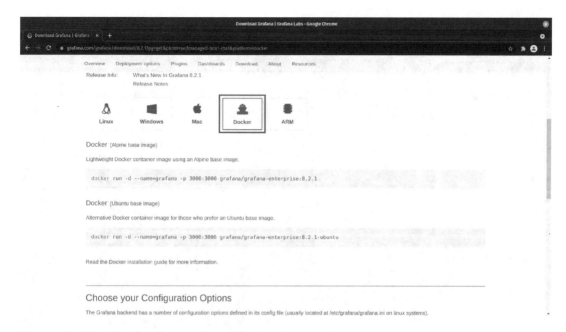

Figure 3-30. *Docker image installation instructions*

To install the Docker image, copy the Docker command for the image you've chosen and paste it into a terminal. For example, to download Grafana Enterprise 8.2.1 using the Alpine base, you would enter the following:

```
docker run -d --name=grafana -p 3000:3000 grafana/grafana-enterprise:8.2.1
```

Note   The -p 3000:3000 section of this command tells Docker to map port 3000 on the host machine to port 3000 of the container image. Without this, you wouldn't be able to access your running Grafana. You can also change this port if you want to use something other than port 3000; consult the docker run documentation for more details.

If this is the first time you've installed a Grafana docker image, you'll see several downloads run before the final image is set up and ready, as shown in Figure 3-31. This is a normal part of the Docker installation process, and the extra images will be cached locally so any later downloads (e.g., if you update the version of Grafana you are using) this process will go faster.

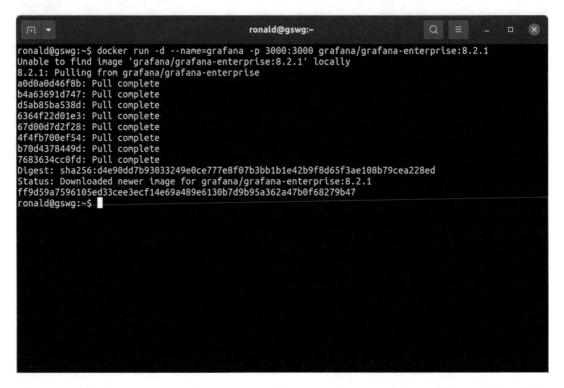

***Figure 3-31.** Running Grafana in a Docker container automatically installs all necessary images*

At this point, you can access Grafana by pointing your web browser to *http://localhost:3000* as shown in Figure 3-32. When Grafana is first started, there's a single administrator user created. The default username and password for this user are both *admin* (all lowercase).

Figure 3-32. *Logging in to Grafana running in a Docker container*

The first time you log in with the administrator account, you'll be prompted to change your password. Be sure to choose a secure password for this account as it has complete control of your Grafana environment.

Once you've changed your password, you're ready to start using Grafana! You'll probably want to skip ahead to Chapter 4 to start connecting data sources to make use of your new Grafana installation.

Raspberry Pi

The Raspberry Pi is a tiny, affordable Linux computer that has become wildly popular in education and for hobbyists looking to add the power of a Linux system to projects without the size or expense of running a full-sized server. The Raspberry Pi is a versatile little system and is a great choice for running a small Grafana environment.

There are a number of different operating systems and environments available for the Raspberry Pi, including various flavors of Linux. Unlike most other desktops or servers in wide use today, which use an Intel (or compatible) processor, the Raspberry Pi is based on ARM processor. (These are the same chips used in most cell phones and tablets.) As a result, the default Linux downloads won't run on a Raspberry Pi.

Grafana Labs provides releases of Grafana built for ARM architectures running Linux, which suits the Raspberry Pi perfectly. For this section, we'll assume you're running Raspberry Pi OS (formerly known as Raspbian), the default version of Linux for the Raspberry Pi.

To download Grafana for your Raspberry Pi, point your browser to *www.grafana.com* and click the "download" link at the top of the page. (You can also go directly to *www.grafana.com/get*.) By default, you'll see information about using Grafana Cloud, so click the "self-managed" tab to see the links to download Grafana directly. Figure 3-33 shows the download page on a Raspberry Pi OS desktop, highlighting the "Download Grafana" link.

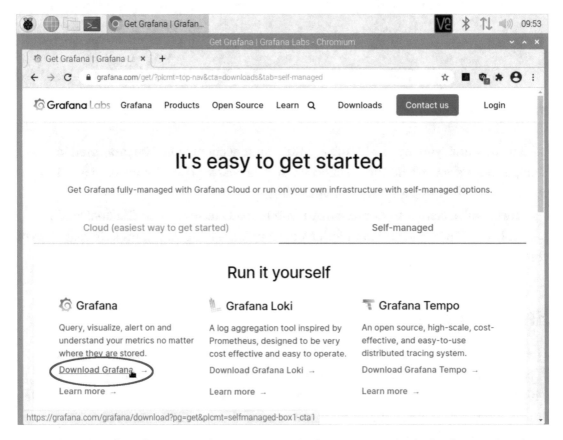

Figure 3-33. *The Chromium browser on Raspberry Pi OS with the "Download Grafana" link highlighted*

After clicking the download link, you'll see the various download options available. Be sure the ARM version is selected by clicking the "ARM" button as shown in Figure 3-34. By default, the latest released version of Grafana will be selected. If you need another version (such as an older release for compatibility, or you want to try a beta version), you can select it here.

Grafana Labs provides an image for both pure open source Grafana and Grafana Enterprise, so you can select either edition to download here.

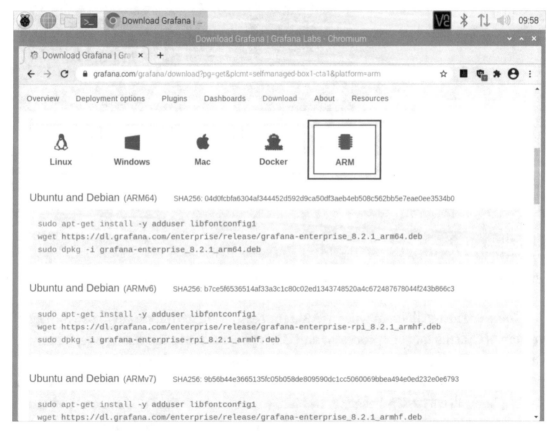

Figure 3-34. *The Grafana download page highlighting the ARM version*

Once you've selected your Grafana version and edition, you'll see some commands that you can copy and paste. (You may need to scroll down to see them.) In order to run these, you'll need to open the terminal application. The terminal icon is in the menu bar at the top of the screen, as highlighted in Figure 3-35. Click this to open a terminal window.

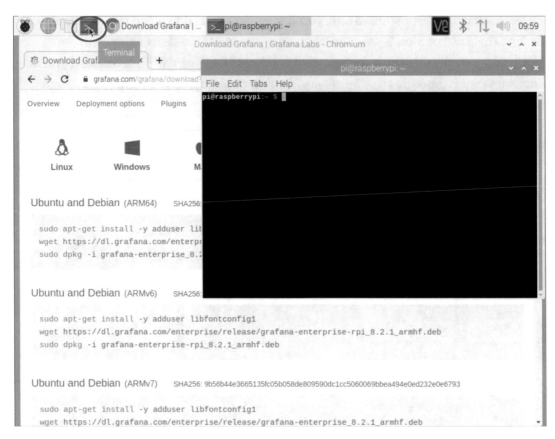

Figure 3-35. *The terminal window launcher and an open terminal*

There are many different types of ARM processor, and it's important to use the right version of Grafana for your processor and OS. If you download the wrong one, it may run slower than expected or might not run at all. Fortunately, all Raspberry Pi devices with the default Raspberry Pi OS run as 32-bit ARMv7 devices, making the choice simple. As Raspberry Pi OS itself is based on Debian, you should use the "Ubuntu and Debian (ARMv7)" version of Grafana. Figure 3-36 shows the correct commands copied and pasted into the terminal.

Note   Some Raspberry Pi devices have other ARM architectures, including 64-bit support. If you are running one of these and know that you are using a 64-bit OS, download the ARM64 version of Grafana. If you're not sure, stick with ARMv7.

Figure 3-36. *The installation commands for Grafana on ARMv7 highlighted and pasted into a terminal*

Listing 3-9 shows these commands in more detail. We'll step through them line by line to understand what's happening.

Listing 3-9. Installing Grafana in a Raspberry Pi OS

```
sudo apt-get install -y adduser libfontconfig1
wget https://dl.grafana.com/enterprise/release/grafana-enterprise_8.2.1_
armhf.deb
sudo dpkg -i grafana-enterprise_8.2.1_armhf.deb
```

> **Note**   You may be prompted for a password after the first command, due to it running `sudo` to temporarily get escalated permissions. Just enter the password that you use to log in to the system here.

While Grafana tries to limit dependencies, there are still a few things that it needs to function correctly that might not be installed on all systems. The first command installs the `adduser` and `libfontconfig1` packages:

```
sudo apt-get install -y adduser libfontconfig1
```

These allow the Grafana package to set up a service account for Grafana to run as for security purposes and to install and configure fonts for rendering data. They may already be installed on your system, in which case this will just verify that and make no other changes.

Next is the actual download of Grafana:

```
wget https://dl.grafana.com/enterprise/release/grafana-enterprise_8.2.1_
armhf.deb
```

This command fetches the installation package for the Enterprise edition of Grafana 8.2.1.

Finally, we install the package using the Debian package manager, dpkg:

```
sudo dpkg -i grafana-enterprise_8.2.1_armhf.deb
```

You should see some informational output after each of these commands. If there are errors, check your Internet connection and carefully read the error messages to troubleshoot the issues.

Once you've run the preceding commands, Grafana is installed! But it still isn't up and running yet. By default, Grafana does not start up automatically, as it's assumed that you might want to make some changes to its configuration to better suit your environment before using it. So we need to tell the Linux system to treat Grafana as a system service and to start it up so that we can use it.

Fortunately, the Grafana package tells us how to do this after installation. Listing 3-10 shows the relevant commands that the installer provides.

Listing 3-10. Enabling Grafana as a service and starting it up

```
sudo /bin/systemctl daemon-reload
sudo /bin/systemctl enable grafana-server
sudo /bin/systemctl start grafana-server
```

These commands work with `systemd`, a commonly used service management system for Linux. Grafana installs itself as a `systemd` service, so the first command tells `systemd` to reload its list of services so that it sees Grafana. Once `systemd` knows that the Grafana service is installed, we can tell `systemd` to `enable` it, which means to set it to start up whenever the system is booted. Of course, we don't really want to have to reboot the system just to start up a new service. So the last command is used to `start` the Grafana service and monitor it to be sure that it will stay up and running. Figure 3-37 shows the output of running these commands successfully.

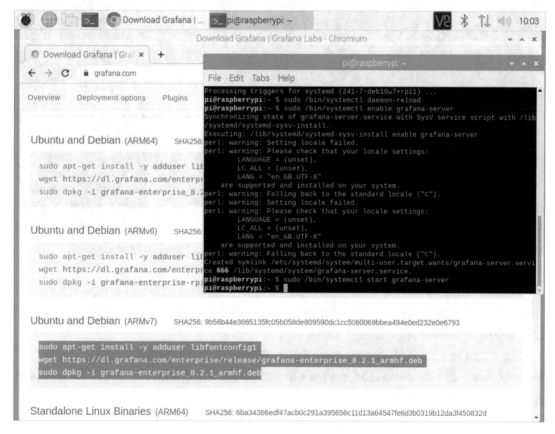

Figure 3-37. *Grafana set up as a service on Raspberry Pi OS*

Now that Grafana is installed and started as a service, you can open a web browser and log in, as shown in Figure 3-38. The address you will use is *http://localhost:3000*. When Grafana is first started, there's a single administrator user created. The default username and password for this user are both *admin* (all lowercase).

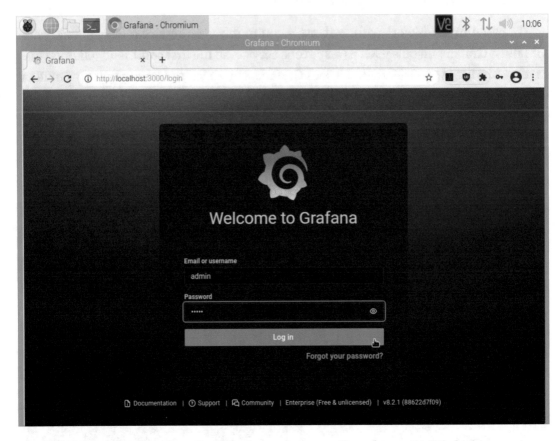

Figure 3-38. *Logging in to Grafana running on a Raspberry Pi OS desktop*

The first time you log in with the administrator account, you'll be prompted to change your password. Be sure to choose a secure password for this account as it has complete control of your Grafana environment.

Once you've changed your password, you're ready to start using Grafana! You'll probably want to skip ahead to Chapter 4 to start connecting data sources to make use of your new Grafana installation.

Summary

In this chapter, you learned how to deploy Grafana in your local environment. (Maybe even more than one environment!) You found the correct version of Grafana for your system, downloaded it, and deployed it. You learned how to start Grafana and log in as the administrator account.

In the next chapter, we'll start adding links to data sources into Grafana. You'll be able to use real data to start building dashboards and visualizing whatever you like.

CHAPTER 4

Connecting to Data Sources

In the previous chapters, we've looked at how to use panels and build dashboards with sample data. While this can be fun, it's not really particularly useful. What we really want to do is visualize real data.

Since Grafana doesn't actually store your data directly, to visualize real data we need to connect Grafana to the *data sources* where that data resides. A data source in Grafana is a plugin that provides a connection to the application or service that contains the data.

It's important to note that data sources *must* provide a query interface in order for Grafana to retrieve data. It has to provide some sort of API or interface that lets you ask for the specific data you want. If there's no way to request data from your system directly, there's no way for Grafana to get it and thus no way for Grafana to display it. What that query interface looks like is up to the data source – as long as it exists and can provide a result when queried, Grafana can work with it. (At least in theory; someone still has to go through the work of writing the plugin before you can use it.) The type of query it supports can be complex and expressive, like SQL, or much simpler, like requesting a specific cell or range from a Google Sheet spreadsheet. But the mechanism for running the query must exist for the data to be used in Grafana.

Tip   This is something a lot of new Grafana users are confused by. People often have data in text or CSV files and want to visualize this in Grafana. But since files on a disk don't provide a query interface, this doesn't work. In order to visualize this data, it first needs to be loaded into a data source such as a SQL database, an online spreadsheet system, or some other service that provides a query interface.

© Ronald McCollam 2022
R. McCollam, *Getting Started with Grafana*, https://doi.org/10.1007/978-1-4842-8309-7_4

A data source plugin gives you a way to query your data inside of Grafana, and how that looks varies depending on the plugin and what it's connecting to. For example, some plugins provide a GUI query builder where you can point and click to select what data you are interested in, while others provide a text-based approach for you to write queries in SQL or other languages. A few even provide multiple modes to make simple queries fast and easy but still let you customize the query where you need more control.

Data sources also format the results of your query in a way that Grafana can understand. Grafana expects all data to be formatted in a single, specific way regardless of how the data is actually stored or formatted originally. So the data source plugin has to take the source data and manipulate it so that it matches the format that Grafana expects. If this didn't happen, every single panel type would need to be able to process data from every possible source. Every time you wanted to connect a new data source to Grafana, you'd have to update the code for every single possible visualization – as you can imagine, this would be a lot of work!

Note   The internal format Grafana uses to represent data is called a *data frame*, which is based on the Apache Arrow project. There's more information about it in the Grafana developer documentation at `https://grafana.com/docs/grafana/latest/developers/plugins/data-frames/`.

You don't need to know anything about data frames unless you want to build a new data source plugin yourself, but if you've ever wondered how Grafana manages to work with so many different types of data storage systems, this is the secret!

In this chapter, we'll look at how to manage data sources in Grafana and configure several types of data sources:

- **Time series databases:** InfluxDB, Prometheus, Graphite

- **Relational databases:** MySQL, PostgreSQL

- **Logs:** Loki, Elasticsearch

We'll also see how to find plugins for Grafana to add support for other data sources.

Managing Data Sources

Data sources are managed in the Grafana configuration panel. This view allows you to add, remove, or update connections to data sources. Figure 4-1 shows the configuration menu open and the data sources option selected and the actual data source list in the background.

Figure 4-1.   *The data source configuration view*

Each entry in the data source configuration view represents one connection to an application or service. It's entirely possible – even common! – to have multiple connections to the same data source. For example, you might have more than one SQL database on a single server. You only need to install the data source plugin once, and then you can create as many connections to that data source as you like.

All of the data source connections are added the same way, by clicking "Add data source" in the configuration view and selecting it from the list of available data sources as shown in Figure 4-2. For environments with a lot of data source plugins loaded, this list can get very long, so using the search box at the top can come in handy.

Figure 4-2. *Adding a data source*

A connection has to be configured before it can be used. Every data source connection has a name, which is what you'll see in Grafana when querying data from it. The rest of the configuration needed is different for each data source, but usually includes things like the location or URL of the data source, usernames, and passwords and sometimes things like database names or required API keys.

It's best practice to give data source connections meaningful names. This is especially true if you have more than one data source connection to the same source! If you only have a single database, "MySQL" might be okay, but it's no fun looking through "MySQL-1," "MySQL-2," etc. to figure out which one has the data you need.

Grafana ships with support for a number of popular data sources already installed. We'll look at how to configure and use some of the most commonly used data sources as follows.

Caution   Most of the following examples assume that you already have a functioning instance of the data source you're configuring and that it's accessible from the Internet (or at least from your Grafana deployment). In some cases, there are sample data sources you can use or free cloud accounts that you can set up to try these services out. While there are links to those where available, it's beyond the scope of this book to explain how to set up and configure data sources external to Grafana.

Testdata DB

Testdata DB is a special case; while it uses the Grafana data source APIs, it's not really a data source at all. Like the "Grafana" data source you used in Chapter 1 when building your first dashboard, Testdata DB generates sample data without actually connecting to a data source.

Where Testdata DB differs from the default sample data source is that instead of being a built-in part of the Grafana interface, it actually provides data through the same data source system that real data sources use. This means that it's much more configurable than the default sample data in Grafana and so can give a much better view of what your real data will look like when you're experimenting.

(In fact, many of the screenshots in Chapter 2 were created with Testdata DB as the default sample data source is too simple to provide meaningful results with several of the panels shown!)

As Figure 4-3 shows, adding a Testdata DB data source is as simple as it gets. You just need to provide a name for your data source, and you can start generating sample data.

Figure 4-3. *Configuring a Testdata DB connection*

Once you've configured the connection, you can use it to generate more interesting sample data in panels. Figure 4-4 shows the default "Random walk" scenario with the options that it provides. Note that unlike the default sample data generator, Testdata DB allows you to do things like generate multiple separate time series in the same graph or set minimums and maximums for your data. This can come in handy if you're mocking up a dashboard that should have data within a specific range.

Figure 4-4. *The default random walk scenario in the Testdata DB data source*

When using the Testdata DB data source in a panel, you'll see a list of different methods for generating data in the *scenario* dropdown menu. Some of these provide non-time series data, so they won't work with the time series panel but will work with other visualizations. For example, the "USA generated data" can be used with the Geomap panel to show geographic distribution of data, as shown in Figure 4-5.

Figure 4-5. *USA generated data from Testdata DB shown in the Geomap panel*

Note   Something interesting you might notice in the scenario dropdown are
entries for "CSV file" and "CSV content." While these can, in fact, be used to
visualize CSV data, they don't dynamically update this data from a user-selected
file or location. The CSV file scenario uses a set of predefined files stored in the
Grafana environment, and the CSV content scenario uses static CSV content that
has been pasted into the plugin.

So while you can technically display CSV data, this doesn't change the note in
the introduction to this chapter about not being able to load and visualize CSV
files directly in the Grafana UI. Unless your data never changes, you still need
something more than Testdata DB to view it.

InfluxDB

InfluxDB is an open source time series database and query engine. It's commonly used for monitoring physical or virtual computer systems as well as sensors or IoT devices. InfluxDB has been around since 2013 and is often used with the Telegraf agent. Telegraf supports over 200 different plugins for collecting data from a wide array of software and hardware, making it a great way to get started with monitoring generally.

Like Grafana, InfluxDB is provided both as a piece of standalone software that can be run locally and as a service that can be used online. A great way to get started with Influx is to sign up for the free cloud tier on the InfluxData website, *www.influxdata.com/*. The cloud-hosted InfluxDB there has some tutorials that will walk you through deploying Telegraf and collecting data from your local system. If you just want to test your Influx connection, there's also some sample data that can be queried without having to deploy Telegraf or another data collector, which is listed at *https://docs.influxdata.com/influxdb/cloud/reference/sample-data/*.

Caution   The hosted InfluxDB service has some continuously updated sample databases, which are called *buckets* in InfluxDB terms. InfluxDB has restricted access to these buckets, so they can only be used within the InfluxDB interface and are not available to use with Grafana.

Any data that you send into the InfluxDB cloud will be available! This restriction is only on the sample data provided by InfluxData.

To set up a connection to an InfluxDB data source in Grafana, start by adding a data source and searching for the InfluxDB plugin, as shown in Figure 4-6.

Figure 4-6. *The InfluxDB data source plugin*

Once you've selected the plugin, the first choice you'll need to make is how you want to query the data. InfluxDB supports two different languages for querying data: *Flux* and *InfluxQL*. InfluxQL is the original query language for InfluxDB and looks very similar to SQL. If you're familiar with SQL and want to use that mode, select InfluxQL in the "Query Language" section of the data source configuration. Flux is a newer language that provides more functionality and power, but does not use a SQL syntax. If you want to use Flux for your InfluxDB queries, choose Flux in the "Query Language" section.

Tip   If you want to mix and match Flux and InfluxQL against the same InfluxDB bucket, you'll need to create two data sources in Grafana, as the InfluxDB plugin for Grafana doesn't support changing these languages on the fly. If this is the case, you might want to call your data sources something like "InfluxDB – Flux" and "InfluxDB – InfluxQL" or similar so that you remember which to use for each query.

You'll also need to add some information to tell Grafana where your InfluxDB server is located and how to authenticate against it. This can vary a bit depending on whether you have chosen Flux or InfluxQL as well as if you are running locally or using the cloud InfluxDB service. We'll look at two configurations as examples.

InfluxQL on Local InfluxDB

If you are running InfluxDB locally, use the URL that you've configured for your InfluxDB server as the URL for the Grafana plugin connection. This URL defaults to *http://localhost:8086* but may be different in your environment. Make sure this is accessible by your Grafana instance – if you're running InfluxDB locally behind a firewall, you'll need to make some changes to your network configuration to access it from Grafana Cloud. (And you'll need a hostname or IP address other than "localhost" which only works locally.)

Unless you've made changes to your InfluxDB environment that require additional configuration, you can leave the rest of the connection options set to their default values, as shown in Figure 4-7. Only the URL field here is required.

Figure 4-7. *Providing a URL for the InfluxDB data source*

You'll also need to tell Grafana how to authenticate against your InfluxDB server. If you're running InfluxDB yourself locally, you'll likely use username/password credentials which you configured when setting up InfluxDB. Figure 4-8 shows adding the InfluxDB database name, username, and password in the "InfluxDB Details" section.

Figure 4-8. *Completing a local InfluxDB connection*

Flux on InfluxDB Cloud

The cloud-hosted InfluxDB uses a more advanced authentication system than a typical local server, so there are a few extra steps to go through. To make things easy, be sure that you're logged in to your InfluxDB cloud environment in another tab or window before you start setting up the data source in Grafana.

To start, there are several items you'll need to get from your InfluxDB account. Most of these are available from the "About" page for your InfluxDB organization. Figure 4-9 shows where to find these items in the InfluxDB web UI.

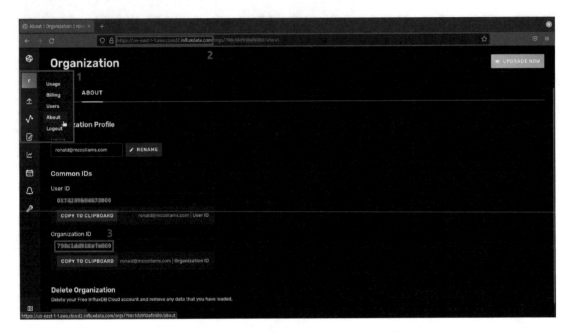

Figure 4-9. *The InfluxDB Organization view*

Note   These instructions assume you are using InfluxDB Cloud 1.x. If you are using a newer version of InfluxDB Cloud, the menu items and views may be slightly different.

Start by clicking your user icon near the upper left and selecting "About," marked as "1" in Figure 4-9. This will take you to a page that has most of the info that you need. You'll first want to copy the beginning of the URL, everything up to but not including the first slash after "influxdata.com," marked as "2" in Figure 4-9. In this example, that would be *https://us-east-1-1.aws.cloud2.influxdata.com*. Paste this into the "URL" field in the Grafana data source plugin configuration as shown in Figure 4-10.

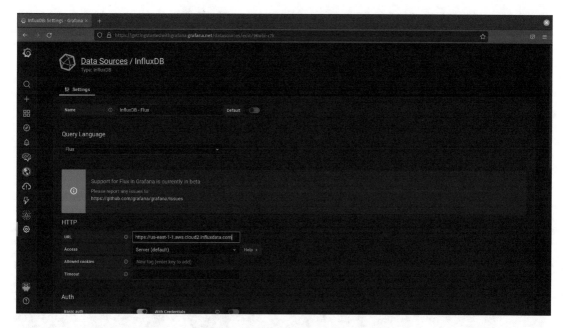

Figure 4-10. *Setting the InfluxDB cloud URL in Grafana*

Finally, copy the "Organization ID" from the InfluxDB site, marked as "3" in Figure 4-9. Paste this into the "Organization" field under "InfluxDB Details" near the bottom of the Grafana data source configuration screen as shown in Figure 4-11.

Figure 4-11. *Setting the InfluxDB Organization ID*

The Organization ID is a sort of username for connecting to InfluxDB, but we still need to configure a secret to be able to access our data. To do this, we'll need to create an API key in the InfluxDB cloud that allows Grafana to access our data.

Figure 4-12 shows the API Tokens control panel in the InfluxDB cloud. Select the Load Data menu on the left (which looks like an arrow pointing up) and then choose the "API Tokens" option. Depending on your use of InfluxDB in the past, you may have one or more API tokens already. Regardless, it's best practice to create a new one specifically for Grafana. To do this, click the "Generate API token" button and select "Read/Write API Token."

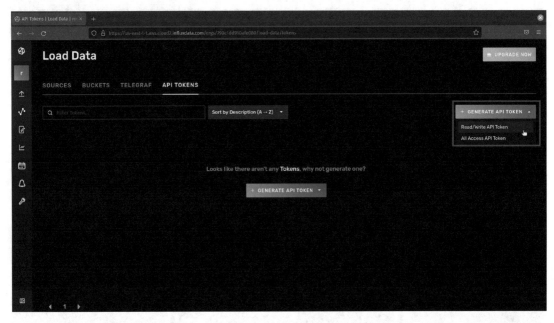

Figure 4-12. *The "Load Data" view in the InfluxDB cloud*

InfluxDB allows you to limit the access that an API token has, shown in Figure 4-13. As a visualization tool, Grafana does not need access to control or write data, so you can limit your API token to only allow reading by clicking "deselect all" under the "Write" section of the API token configuration window. If you want to restrict the token further, you can select specific buckets for this key. Selecting "all buckets" will allow this key to read any data in your InfluxDB account. When you've set the options the way you like, click "save."

Figure 4-13. *Creating an InfluxDB API token*

Once you've saved the token, you can rename it to something memorable by clicking the pencil icon. Figure 4-14 shows the token itself, brought up by clicking the name of the token. Copy this into the "Token" field in the Grafana data source configuration view as shown in Figure 4-11.

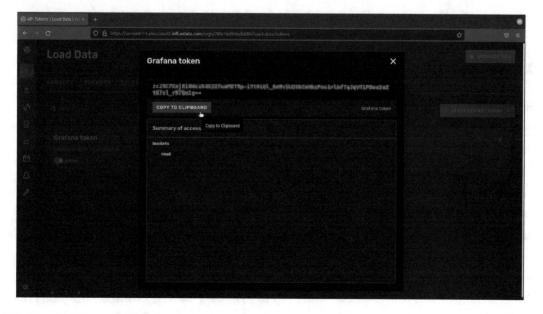

Figure 4-14. *An InfluxDB token*

Wrapping Up

Once you've configured your options, click the "Save & test" button in Grafana. If everything is configured correctly, you'll be able to use the data source in panels now. Figure 4-15 shows an InfluxQL query showing some server load data.

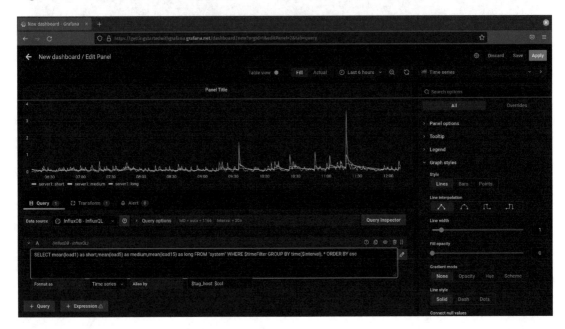

Figure 4-15. *Querying InfluxDB with InfluxQL*

Prometheus

Prometheus is a popular open source metric store, created in 2012 at SoundCloud and moved to the Cloud Native Computing Foundation in 2016. Since then, Prometheus has steadily increased in popularity due both to being easy to deploy and being the default metric system for Kubernetes, a widely used container orchestration system.

Unlike many other metric collection systems which run an agent on monitored devices and push data to a central server, by default Prometheus uses a pull model: a central Prometheus server is configured with locations where it can find data, and it pulls (or *scrapes* in Prometheus language) the metrics from the systems being monitored. Fortunately, if you don't want to run your own Prometheus infrastructure, you can use Grafana Cloud and a tool called Grafana Agent to collect data in a more traditional push model. The Grafana Agent can be found at `https://github.com/grafana/agent/` and is available for many common environments.

As a core data source for Grafana, adding a connection to Prometheus is straightforward. In fact, if you're using Grafana Cloud, you have a Prometheus data source already set up and configured. You can download the Grafana Agent and start pushing data into the cloud.

Let's take a look at configuring a Prometheus data source in Grafana. This might be an on-premise deployment of Prometheus or could be used to connect to the Prometheus endpoint of Grafana Cloud Metrics from a locally deployed Grafana instance.

Start by adding a data source and searching for the Prometheus data source as shown in Figure 4-16.

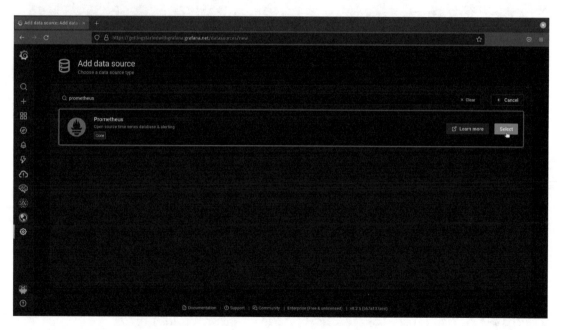

Figure 4-16. *Adding a Prometheus data source to Grafana*

Prometheus exposes its metric query interface via HTTP, by default on port 9090. So if you are running Prometheus and Grafana on the same server, you can likely access Prometheus via *http://localhost:9090*. If they're not on the same server, remember that Grafana needs to be able to access the Prometheus instance over the network in order to display data. So if you are running Prometheus inside your firewall but want to run Grafana in the cloud, you'll need to configure your network to allow this.

For most standard Prometheus environments, the URL is the only thing you need to configure. Standard Prometheus does not have any notion of per-user security, so there are no usernames or passwords to configure. So let's take a look at what you need to do for a slightly more complex environment – we'll set up a connection to a Grafana Cloud Metrics Prometheus environment.

To begin, you'll need to log in to your Grafana Cloud account at *https://grafana. com/*. Once logged in, click "My account" to see your cloud configuration. Figure 4-17 shows where you will find your Prometheus information. Locate this box and click the "Details" button.

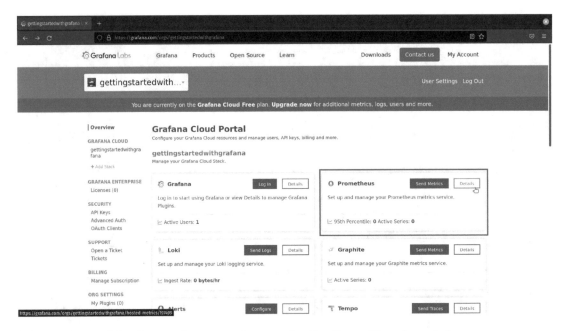

Figure 4-17. *The Prometheus section on the Grafana Cloud portal*

Clicking through to the Prometheus section of the cloud portal will give you a wealth of detail about your Prometheus instance. There are instructions here for sending data into the cloud in various ways, including with the Grafana Agent. If you want to have some example data to play with, follow the instructions here to deploy the Agent, and you will start sending metrics from your local computer to the cloud.

To configure a Prometheus connection in Grafana, though, we need to look for not only a URL but also a username and password. Your username, or instance ID, is generated automatically and will be visible on the Prometheus page. But you'll need to generate a password, or API key, to use it to connect. Figure 4-18 shows the URL, instance ID, and API key sections of the cloud portal page. Start by copying the URL from the query endpoint section into the URL field of your Grafana Prometheus data source configuration.

Figure 4-18. The Grafana Cloud Prometheus instance ID and API key section

---

Caution   Be sure to copy the "query endpoint" URL and not the "remote write endpoint" URL. The remote write URL is used to send data into Grafana rather than query it.

In order to see the fields for username and password in the Grafana Prometheus data source, you'll need to select the "Basic auth" option. Once that's done, you can copy the instance ID into the username field in Grafana.

The last step required is to create an API key which will function as a password. You'll see a link titled "generate now" in the API key field on the Grafana Cloud portal. Click that to create a new key, as shown in Figure 4-19. Give the key a memorable name and set its role to "Viewer" so that it has permission to read data in Prometheus, and click the "Create API key" button.

Figure 4-19. *Creating a Prometheus API key in Grafana Cloud*

Once you click this button, you'll be shown the API key that was generated. Copy this key and paste it into your Grafana data source configuration as the password. Be sure to do this right away – as the key page tells you, this key will only be displayed once and then never again!

Once you've added the username and password, click the "Save & test" button in Grafana to finish setting up the data source. If you get an error, check to be sure everything is filled out correctly.

Once set up, you will be able to query Prometheus data! (Though remember that if you're using the Prometheus data source in Grafana Cloud Metrics, there won't be any data in there by default. You'll still need to send some data using the Grafana Agent or some other method before you can query anything.)

Graphite

Graphite is older than InfluxDB and Prometheus, the other two time series databases in this chapter, dating to 2006 at Orbitz. But while these newer systems may generate more buzz now, Graphite is still widely used and is in production at some of the largest Internet and ecommerce services.

Graphite is a core Grafana plugin, meaning that it's included by default and easy to configure without adding anything extra to Grafana. Grafana Cloud Metrics provides a Graphite endpoint by default, so if you have created a Grafana Cloud account already, you already have access to a Graphite environment and can start visualizing data in Grafana.

Let's take a look at setting up Graphite as a data source in Grafana. This might be an on-premise deployment of Graphite or could be used to connect to the Graphite endpoint of Grafana Cloud Metrics from a locally deployed Grafana instance. Start by adding a data source and searching for the Graphite data source as shown in Figure 4-20.

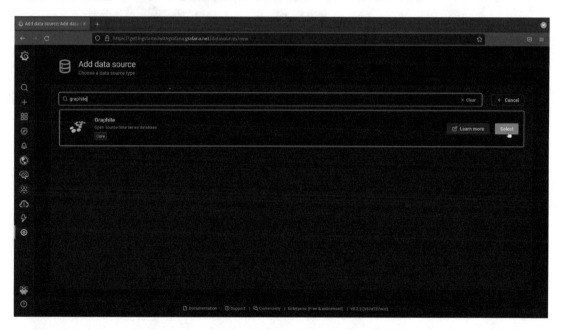

Figure 4-20. *Adding a Graphite data source to Grafana*

Graphite runs its own web server to expose metrics, usually on port 8080 or port 80. So if you are running Graphite and Grafana on the same server, you can likely access Graphite via *http://localhost:8080/* or *http://localhost/*. If they're not on the same server, remember that Grafana needs to be able to access the Graphite instance over the network in order to display data. So if you are running Graphite inside your firewall but want to run Grafana in the cloud, you'll need to configure your network to allow this.

In many cases, you can connect directly to a Graphite data source with just the URL that you use to view Graphite directly. In simple deployments where you do not need a username and password to view data in Graphite, just copying the URL into the URL field in the Grafana data source configuration page will be all that you need.

In this example, let's look at setting up a connection that requires credentials to access. For this, we'll set up a connection to a Grafana Cloud Metrics Graphite environment.

To begin, you'll need to log in to your Grafana Cloud account at *https://grafana. com/*. Once logged in, click "My account" to see your cloud configuration. Figure 4-21 shows where you will find your Graphite information. Locate this box and click the "Details" button.

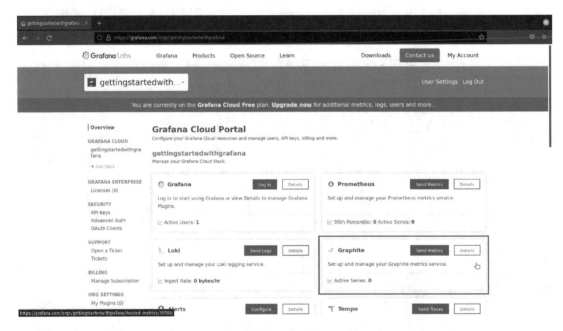

Figure 4-21.   *The Graphite section of the Grafana Cloud portal*

Clicking through to the Graphite section of the cloud portal will show you relevant information about your Grafana Cloud Metrics environment. There are instructions here for sending data into the cloud in various ways. These instructions assume that you already have some local Graphite agents set up and collecting data. If you don't have this configured already, you'll need to visit the Graphite website at *https://graphiteapp. org/* to download and configure the local infrastructure first.

To configure a Grafana Cloud Metrics Graphite connection in Grafana, we need not only a URL but also a username and password. Your username, or instance ID, is generated automatically and will be visible on the Graphite page. But you'll need to generate a password, or API key, to use it to connect. Figure 4-22 shows the URL, instance ID, and API key sections of the cloud portal page. Start by copying the URL from the query endpoint section into the URL field of your Grafana Graphite data source configuration.

Figure 4-22. *The data source settings for Graphite hosted on Grafana Cloud*

In order to see the fields for username and password in the Grafana Graphite data source, you'll need to select the "Basic auth" option. Once that's done, you can copy the instance ID into the username field in Grafana.

The last step required is to create an API key which will function as a password. You'll see a link titled "generate now" in the API key field on the Grafana Cloud portal. Click that to create a new key, as shown in Figure 4-23. Give the key a memorable name and set its role to "Viewer" so that it has permission to read data in Graphite, and click the "Create API key" button.

Figure 4-23. *Creating a Graphite API key*

Once you click this button, you'll be shown the API key that was generated. Copy this key and paste it into your Grafana data source configuration as the password. Be sure to do this right away – as the key page tells you, this key will only be displayed once and then never again!

Once you've added the username and password, click the "Save & test" button in Grafana to finish setting up the data source. If you get an error, check to be sure everything is filled out correctly.

Once set up, you will be able to query Graphite data! (Though remember that if you're using the Graphite data source in Grafana Cloud Metrics, there won't be any data in there by default. You'll still need to send some data using a supported Graphite agent or relay.)

MySQL

MySQL is one of the most commonly used open source SQL databases, popular among web developers for its simplicity of deployment and management. If a web host offers only a single database option, odds are MySQL will be it. This means that for non-time series data like content storage, product sales, or other business data, it's a natural fit. Being able to display this data alongside of monitoring and logging data in Grafana lets you look at your whole environment in one place.

Because MySQL is a core plugin for Grafana, you won't need to install anything extra to be able to connect to it and start visualizing your data. Unlike the time series databases covered earlier in the chapter, MySQL does not have its own built-in web interface. Instead, it listens for connections using its own protocol on port 3306. So unless you've changed your MySQL configuration, you'll need to be sure that port is accessible from your Grafana server. If they are running on the same system, you can probably use *localhost:3306* to connect, but if your Grafana instance is outside your firewall (e.g., in Grafana Cloud), you'll need to make the necessary network configuration changes to allow inbound access to your MySQL environment.

Tip   This additional network configuration applies to cloud providers or web hosts as well. You'll frequently find that a hosted MySQL service is only available from other parts of the hosted environment by default, even if your web server is publicly available. This includes public cloud providers' hosted MySQL as well. Check the documentation for your hosting service to learn how to open inbound access to your MySQL instance.

To set up MySQL as a data source in Grafana, start by adding a data source and searching for the MySQL data source as shown in Figure 4-24.

Figure 4-24. *Adding a MySQL data source to Grafana*

For typical MySQL environments, you'll need only three pieces of information: the host (or address of the server that is running MySQL), a username, and a password. There are other options that can be enabled if you have additional security settings turned on in your environment, but these three things should be all you typically need to get a MySQL connection working. Figure 4-25 shows the configuration of the host, user, and password fields for a MySQL environment hosted in Amazon RDS. Note that the port has to be included, even if you're using the default of 3306.

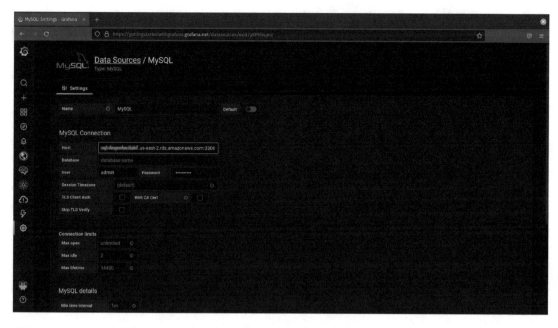

Figure 4-25. *Configuring the MySQL connection*

Once you've configured the host and your user credentials, click the "Save & test" button in Grafana to finish setting up the data source. Check the Grafana UI for errors – in particular, timeout errors likely mean that your MySQL server is not accessible from the Grafana instance.

After you've added the MySQL data source, it will be available for you to use in panels on dashboards. By default, most Grafana visualizations work with time series data, so they expect to see a time or date field somewhere in the data. If your data doesn't have this, you can still use it in Grafana! You'll just need to select a visualization that does not require time series data. Figure 4-26 shows an example of the table visualization showing the example "employee" data set from MySQL. (Note that to make it easier to run custom SQL queries, I've also switched from the graphical query builder to a raw query entry box by clicking the pencil icon above the query.)

Figure 4-26. *Querying tabular data from MySQL*

If you need a refresher on which visualizations are most suited to non-time series data, refer to Chapter 2.

PostgreSQL

PostgreSQL, like MySQL, is a widely used open source SQL database. PostgreSQL has a strong emphasis on correctness and robustness, meaning that while it might take a bit of expertise to configure perfectly, it is ideal for mission-critical data.

Because PostgreSQL is a core plugin for Grafana, you won't need to install anything extra to be able to connect to it and start visualizing your data. PostgreSQL does not provide a built-in web interface like the time series databases we looked at earlier. Instead, it listens for connections using its own protocol on port 5432. So unless you've changed your PostgreSQL configuration, you'll need to be sure that port is accessible from your Grafana server. If they are running on the same system, you can probably use *localhost:5432* to connect, but if your Grafana instance is outside your firewall (e.g., in Grafana Cloud), you'll need to make the necessary network configuration changes to allow inbound access to your PostgreSQL environment.

Tip   This additional network configuration applies to cloud providers or web hosts as well. You'll frequently find that a hosted PostgreSQL service is only available from other parts of the hosted environment by default, even if your web server is publicly available. This includes public cloud providers' hosted PostgreSQL as well. Check the documentation for your hosting service to learn how to open inbound access to your PostgreSQL instance.

Setting up PostgreSQL as a data source in Grafana starts with you adding a data source and searching for the PostgreSQL data source as shown in Figure 4-27.

Figure 4-27. *Adding a PostgreSQL connection to Grafana*

For most PostgreSQL environments, you'll just need to know the host (or address of the server running PostgreSQL), a username, and a password added to the appropriate fields as shown in Figure 4-28. Some environments will require additional security certificates to enable TLS/SSL encryption on top of the connection. If you don't have the certificate data, this is usually safe to ignore – just set the "TLS/SSL Mode" setting to "disable" in Grafana. You'll still have an encrypted connection, but Grafana will not attempt to verify the certificate. If you have the certificate data, you can either copy the certificate and paste it into Grafana (using the "Certificate content" setting) or put it in a file on the Grafana server (using the "File system path" setting). Check with your database administrator or your hosting service documentation for more information about these certificates.

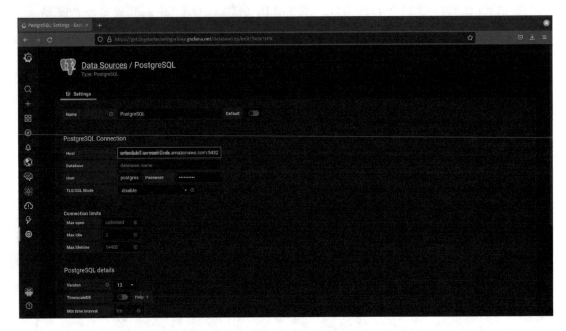

Figure 4-28. *Configuring the PostgreSQL connection*

Caution   Disabling the certificate checking in Grafana means that you'll no longer be notified if the certificate changes or is invalid. This means that if someone gains access to your database server, they could conceivably get access to your data or to usernames and passwords without you being aware. Never disable certificate checking for critical or sensitive data.

Once you've configured the host and your user credentials, click the "Save & test" button in Grafana to finish setting up the data source. Check the Grafana UI for errors – in particular, timeout errors likely mean that your PostgreSQL server is not accessible from the Grafana instance.

After you've added the PostgreSQL data source, it will be available for you to use in panels on dashboards. By default, most Grafana visualizations work with time series data, so they expect to see a time or date field somewhere in the data. If your data doesn't have this, you can still use it in Grafana! You'll just need to select a visualization that does not require time series data. Figure 4-29 shows an example of the table visualization showing tabular data from PostgreSQL, including some of its full text searching capabilities. (Note that to make it easier to run custom SQL queries, I've also switched from the graphical query builder to a raw query entry box by clicking the pencil icon above the query.)

Figure 4-29. *Querying tabular data from PostgreSQL*

If you need a refresher on which visualizations are most suited to non-time series data, refer to Chapter 2.

Loki

Grafana is most known for visualizing time series data, but it's fully capable of querying and displaying log data as well. Loki is a newer open source project, started in 2018, but has quickly grown in popularity due to its lightweight indexing system and close ties to Prometheus. It also uses Grafana as its primary user interface, meaning that Prometheus, Loki, and Grafana all work seamlessly together.

Much like Prometheus, Loki can be run locally or used in a hosted environment. Grafana Cloud provides a hosted Loki instance with a free tier so we'll use that as an example, but setting up the connection in Grafana works the same locally. To start, add a data source and search for Loki as shown in Figure 4-30.

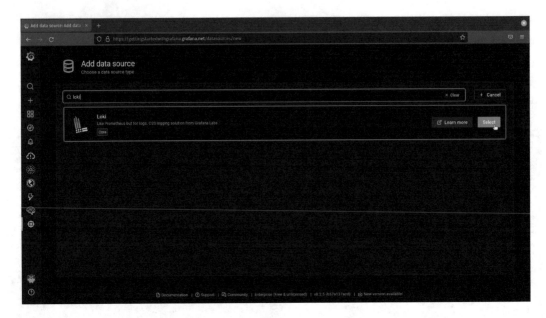

Figure 4-30. *Adding a Loki data source*

To get the connection settings for your hosted Loki instance, you'll need to log in to your Grafana Cloud account at *https://grafana.com/*. Once logged in, click "My account" to see your cloud configuration. Figure 4-31 shows where you will find your Loki information. Locate this box and click the "Details" button.

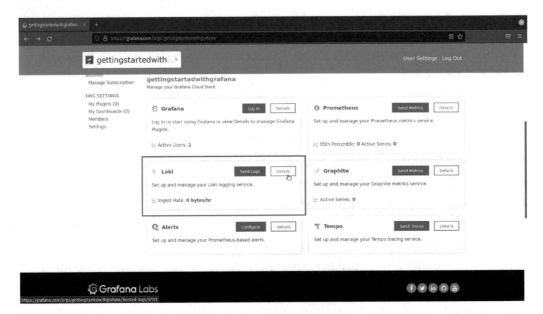

Figure 4-31. *The Loki section of the Grafana Cloud portal*

Clicking through to the Loki section of the cloud portal will show you relevant information about your Grafana Cloud Logs environment. There are instructions here for sending data into the cloud in various ways. Loki uses an agent called *promtail* to collect logs and send them to the Loki server. If you haven't already started sending logs to the cloud, you'll need to set up at least one agent to send in logs – otherwise, you will not see any data once you've connected Loki to Grafana.

To configure a Grafana Cloud Logs Loki connection in Grafana, we need not only a URL but also a username and password. Your username, or instance ID, is generated automatically and will be visible on the Loki page. But you'll need to generate a password, or API key, to use it to connect. Figure 4-32 shows the URL, instance ID, and API key sections of the cloud portal page. Start by copying the URL from the query endpoint section into the URL field of your Grafana Loki data source configuration.

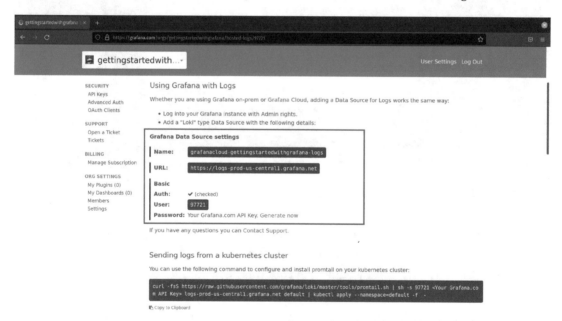

Figure 4-32. *The data source settings for Loki hosted on Grafana Cloud*

In order to see the fields for username and password in the Grafana Graphite data source, you'll need to select the "Basic auth" option. Once that's done, you can copy the instance ID into the username field in Grafana.

The last step required is to create an API key which will function as a password. You'll see a link titled "generate now" in the API key field on the Grafana Cloud portal. Click that to create a new key, as shown in Figure 4-33. Give the key a memorable name and set its role to "Viewer" so that it has permission to read data in Loki, and click the "Create API key" button.

Figure 4-33. *Creating a Loki API key*

Once you click this button, you'll be shown the API key that was generated. Copy this key and paste it into your Grafana data source configuration as the password. Be sure to do this right away – as the key page tells you, this key will only be displayed once and then never again!

Once you've added the username and password, click the "Save & test" button in Grafana to finish setting up the data source. If you get an error, check to be sure everything is filled out correctly.

Once set up, you will be able to query Loki data! (Though remember that if you're using the Loki data source in Grafana Cloud Metrics, there won't be any data in there by default. You'll still need to send some data using a supported promtail.) Figure 4-34 shows a simple Loki query using the "logs" visualization in Grafana, though with the appropriate queries you can use many of the time series visualizations with Loki as well.

Figure 4-34. *Querying Loki logs in Grafana*

Elasticsearch

Elasticsearch is one of the oldest and most widely adopted open source tools for log aggregation and searching. While Elasticsearch is not actually limited to working with just log data, it works extremely well for this use case.

Elasticsearch can be run locally or consumed as a service online. The configuration for these is the same in Grafana, so to make things easy we'll look at using Elasticsearch as a service in this example. Elastic provides a free 14-day trial of Elasticsearch at www.elastic.co along with some sample data that you can import if you don't have data of your own. You can also use tools such as Elastic Beats or Logstash to send in data. Be sure you have some data, either real or sample data, available in Elasticsearch before you try to run queries in Grafana.

Figure 4-35 shows how to start the process in Grafana by adding a data source and searching for Elasticsearch.

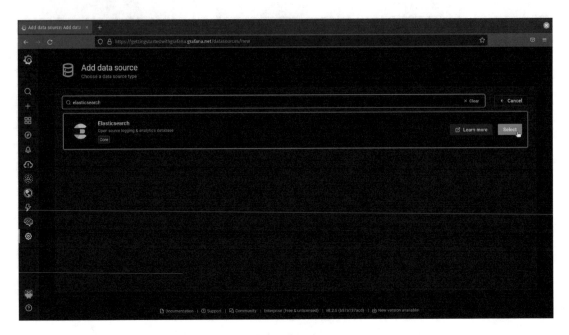

Figure 4-35.   *Adding an Elasticsearch data source*

In order to set up your Elasticsearch data source in Grafana, you'll need three pieces of information: a host (or address of the server that is running your Elasticsearch instance), a username, and a password. If you are running Elasticsearch locally, you will have configured these values when you installed the system, but if you're using the Elastic hosted Elasticsearch environment, these will have been created for you. The administrator name defaults to "elastic," and the password will be displayed only once when the environment is created (or if you later reset it) as shown in Figure 4-36. Be sure to copy it down for safekeeping as you'll need it to connect to Grafana.

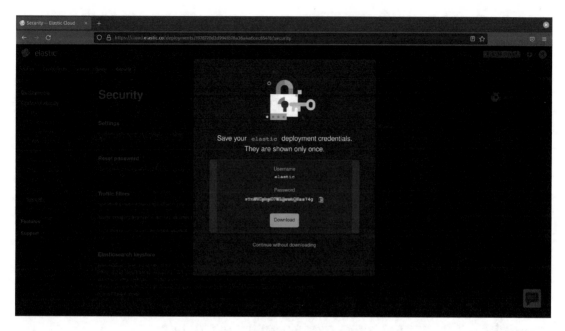

Figure 4-36. *Copying the autogenerated password for the "elastic" account*

In Grafana, select the "basic auth" setting in your Elasticsearch data source and add "elastic" as the user and paste the password from the Elastic cloud into the password field.

The last piece of information you'll need to finish configuring your Elasticsearch data source is the Elasticsearch host. Elasticsearch listens on port 9200 by default, so if you are running Elasticsearch on the same server as Grafana, you can use *http://localhost:9200* for this. If you're running your own Elasticsearch server on another host, be sure that inbound access on this port is allowed from the Grafana server. You may need to make changes to your firewall configuration to allow this access.

If you're using the hosted Elasticsearch instance provided by Elastic, you can find the host information in your hosted deployment settings page. Click the link to copy the endpoint from here as shown in Figure 4-37. Paste this link into the host field in your Grafana data source.

Figure 4-37. *Copying the hosted Elasticsearch host endpoint*

Finally, make sure that the version of Elasticsearch you're using is set properly in your Grafana data source. If you're using a hosted instance from Elastic, this will always be the latest version, but if you're running your own Elasticsearch instance, be sure to set the proper version as shown in Figure 4-38.

Figure 4-38. *Setting the version of Elasticsearch in the Grafana data source*

Once you've finished configuring your Elasticsearch data source, click the "Save & test" button to save your configuration. Take note of any errors; in particular, a timeout error likely means that you do not have access to your Elasticsearch instance from your Grafana server, so check your firewall settings.

Once finished, you can query data from Elasticsearch as metrics, logs, or simply raw tabular data. Figure 4-39 shows an example of an Elasticsearch query run in Grafana, showing log data formatted in the table visualization.

Figure 4-39. *Visualizing log data in Elasticsearch with the table visualization*

Tip   Elasticsearch data can be visualized in two ways in Grafana: as numbers or as text. Numbers are useful for seeing trends like the number of error messages seen per minute. Text is more useful for specific details, like which IP addresses are accessing a resource. If you want to see raw log data, be sure to select the "logs" metric type in your query, as highlighted in Figure 4-39.

Installing Other Data Sources

Grafana ships with more than a dozen "core" data sources which don't need any extra installation or configuration beyond setting up host and authentication details. But there are far more systems that contain data than these core data sources!

Grafana is designed to be extensible through a plugin system. That is, it allows you to add additional functionality by installing plugins that are not part of the default installation, but can be added and removed as needed.

You can find additional data sources to add to Grafana under the "plugins" section of the configuration menu, as shown in Figure 4-40.

Figure 4-40. *The plugins configuration panel showing data source plugins*

By default, the plugins configuration panel will show you all installed plugins. This includes not only data sources but panels (other data visualizations) and applications (extensions to Grafana's core functionality). To install new data sources, use the filters at the top of the panel to show only data sources, and be sure to select all plugins or you'll see only plugins that are already installed.

Note   Some of these plugins will be marked with an "Enterprise" tag, meaning that these are part of the Grafana Enterprise suite and not available for free. We'll look at Grafana Enterprise in more depth in Chapter 13.

For example, if you're a software developer, you might be interested in tracking GitHub metrics on things like number of commits, releases, bugs, etc. You could write a script to look at these metrics and export them to a database, but then you need to worry about making sure your exporting service is always running. But GitHub has an API to get this data, so it's far more convenient to just connect directly to the source!

Figure 4-41 shows the GitHub plugin button in the Grafana configuration panel. Clicking this will allow you to immediately install the plugin. (It may take a minute or two for the plugin to be available once installed.)

Figure 4-41. *Installing the GitHub plugin*

After a data source plugin has been added, you still need to go through the process of setting up the data source connection. The plugin provides the *ability* for Grafana to connect to the data source, but until you create a new data source entry, Grafana doesn't have enough information to actually make that connection.

Once you've done that, you can query your data source like any other, as shown in Figure 4-42. Aside from the fact that core data sources ship with Grafana, there's no other difference between them and other data source plugins. All installed data sources are first-class citizens of the Grafana platform.

Figure 4-42. *The GitHub data source plugin installed and querying the author's repositories*

Summary

In this chapter, we looked at configuring several of the most popular data sources in Grafana, including time series databases for metrics, SQL databases for relational data, and logging systems for visualizing log data. We also saw how to extend Grafana to new data sources by using plugins.

In the next chapter, we'll keep looking at customizing and extending Grafana by seeing how to manage users and groups, set permissions on dashboards and control access to data, and even connect to directory services to have Grafana accounts managed centrally by IT.

CHAPTER 5

User Administration

So far, we've looked at Grafana from the perspective of a single user. That user – you – has full access to everything inside of Grafana. They can add or remove data sources and plugins, change dashboards on a whim, and even delete everything that you've built with a couple of clicks.

Operating Grafana as a single user is fine for a lot of uses, but in shared environments you probably want to have some roles assigned. For example, you might want to limit who has access to certain dashboards. (Even if you don't have security reasons to limit access to data, some dashboards just aren't relevant to certain people, so why clutter up their display and make it harder for them to find what they are looking for?) You might also want to let some users view dashboards but not modify them. (Again, even if you don't need to restrict access for security reasons, it's nice to know that viewers of your dashboards won't change or delete them by mistake!) And of course, it makes sense to limit who can administer the system to add or remove plugins.

Grafana provides some simple mechanisms to create and manage user accounts for these reasons. There are also larger groups of users, dashboards, and settings that can help you run Grafana in a multitenant environment.

In this chapter, we'll look at how to create and manage users and permissions in Grafana, how to organize sets of users and dashboards, and how to connect to directory services.

User Roles

Grafana has three built-in default roles for users: admin, editor, and viewer. There's also a superuser role that allows managing the Grafana instance itself.

© Ronald McCollam 2022
R. McCollam, *Getting Started with Grafana*, https://doi.org/10.1007/978-1-4842-8309-7_5

Each of these roles has a set of permissions that are applied by default unless overridden specifically. For example, a user with the *viewer* role will be able to view but not edit dashboards by default, but can still be given the *editor* role for a specific dashboard:

- The **viewer** role allows viewing of dashboards, but not editing.

- The **editor** role allows full visibility and editing of all dashboards, but not the ability to change data sources.

- The **admin** role allows full visibility and editing of all dashboards, the ability to add, remove, and modify data sources. They can also add users and change user roles.

- The **server admin** role allows full access to all parts of the environment. A server admin can add and remove plugins, view and edit all dashboards, add and remove users, and change existing roles and permissions on any object in the system.

Note   The server admin role is only available in Grafana instances that you deploy and manage yourself. It's not available in Grafana Cloud. Grafana Cloud provides a different way of configuring the underlying Grafana instance, which we'll look at in more detail in the following user management section.

We'll look at how to assign these roles as follows.

Managing Users

In order to manage users in Grafana, you will need to be logged in with an account with the admin role. If you're using Grafana Cloud, this will be the user account you signed up with, as the person who creates the account is automatically made an admin. If you're running Grafana yourself, there is a default account called *admin* that has this role. Either way, you can grant the admin role to other accounts, so there's no need to have a single administrative service account.

> **Tip** If you've forgotten, the default password for the "admin" account is "admin." You should definitely change this to something more secure if you haven't already!

Adding or Removing Users

How you add and remove users depends on whether you're running Grafana yourself or using Grafana Cloud. While Grafana itself stores user account information, Grafana Cloud provides a special directory of user account data to Grafana. This lets you have access to multiple different Grafana Cloud instances without having to have a separate login for each one.

Self-Managed Grafana

To manage users in a Grafana that you've deployed yourself, start by logging in as a user with the admin role. Once you do, you'll see the configuration icon in the menu bar on the left. Choosing the "Users" item in the menu will take you to the user management screen as shown in Figure 5-1.

Figure 5-1. *The user menu item and the user configuration screen in Grafana*

To add a user, click the "Invite" button in the upper-right corner. This will take you to a screen where you will enter information about the user you are adding. An email address is required, and you can optionally add a more friendly name. Figure 5-2 shows the user invitation form filled out to invite a new user to a Grafana instance.

Figure 5-2. *Adding a new user to a self-hosted Grafana instance*

When adding your new user, you can set their role as well. This can be changed later, but it's often most convenient to just configure this at the same time.

The new user will need to click a special link to finish setting up their account. By default, Grafana will send an email to the new user with this link. If you deselect the "Send invite email" option, however, no email will be sent. You can get the link that would be emailed to them from the user configuration screen by clicking "Pending Invites" and using the button there to copy the invitation link, as shown in Figure 5-3.

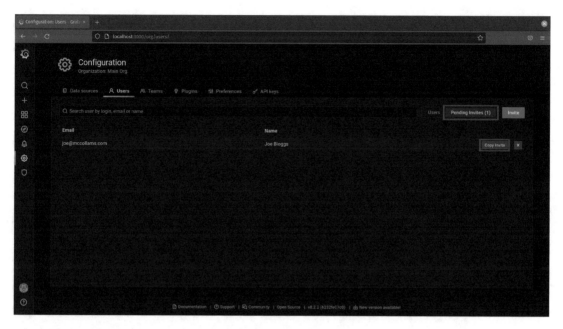

Figure 5-3. *The user configuration screen highlighting the "Pending Invites" view and the "Copy Invite" button*

Tip   If you have deployed Grafana but not made any changes to the configuration file, you won't have an email server configured. This means that email invitations won't be sent out at all, and the only way to add users is to copy the special link from the configuration screen.

You'll need to edit the "smtp" section of your Grafana settings file and restart Grafana to configure this. More information is available in the Grafana documentation at *https://grafana.com/docs/grafana/latest/ administration/configuration/#smtp.*

To delete a user, just click the red X next to their account information in the user configuration screen. Once you do this, you'll be asked for confirmation as shown in Figure 5-4. If you're really sure you want to delete that user, click "Delete" and they'll be removed from your Grafana environment.

Figure 5-4. *Deleting a user from a self-managed Grafana instance*

Grafana Cloud

Grafana Cloud manages users centrally, so you'll need to go through the cloud admin page to add users. The easiest way to get to this page is to start by going to the user configuration screen by choosing "Users" from the menu under the configuration icon in the navigation bar on the left. As you'll see in Figure 5-5, there's a link from here directly to the Grafana Cloud management portal.

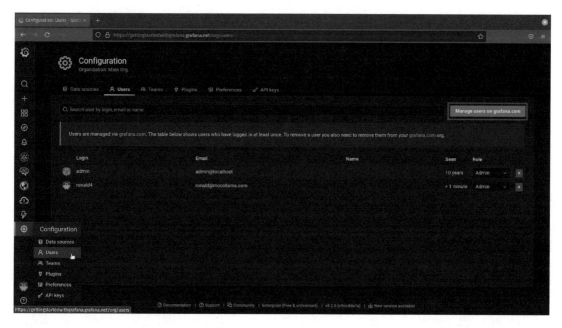

Figure 5-5. *The Grafana user configuration screen with the link to the Grafana Cloud management portal highlighted*

Clicking the link will open the Grafana Cloud management portal in a new browser tab. (Note that you may need to log in to Grafana Cloud if you are not already. If you get a "404 not found" message, check to be sure you're logged in and then try again.) In this screen, you can make changes to existing users in your environment and invite new ones. Figure 5-6 shows the Grafana Cloud user management screen including the button in the upper right to add a new user to your environment.

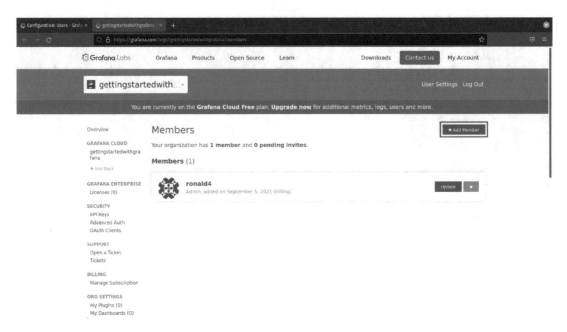

Figure 5-6. *The Grafana Cloud user management screen with the "Add Member" button highlighted*

When you click this button to add a new user to your environment, you'll be asked for a few pieces of information as shown in Figure 5-7. You'll need to fill in an email address which will be used to send a link to the user inviting them to your account. They'll need to click this link in order to be added. You can also set their role here, though you can change it later. Finally, there's an option to have the new user receive billing emails. This only applies for paid Grafana Cloud accounts. If this option is turned on, this user will be emailed a copy of the monthly billing statement for the paid Grafana Cloud service. Once you have filled everything in, click the "Send Invite" button to invite the new user.

Figure 5-7. *Inviting a new user to Grafana Cloud*

Deleting a user in Grafana Cloud works similarly. In the Grafana Cloud admin portal, simply click the red X button next to a user's name to remove them from your environment.

Changing User Roles

User roles (admin, editor, viewer) are managed in the Grafana user configuration screen. To change a user's role, just select the appropriate new role for that user in the dropdown next to their name, and their role will be updated immediately.

Tip   User roles are just sets of default permissions and can be overridden by more specific team or dashboard permissions. So changing someone from an editor to a viewer will remove their ability to make changes to dashboards **except where they have explicit permission already**. Be sure to review the next section in this chapter on permissions to understand how team and dashboard permissions work!

Teams

It's often useful to be able to refer to a group of users all together, especially when organizing dashboards and managing permissions. Grafana calls these groups of users *teams*. Teams are a way of logically organizing users for ease of management.

A user can be a member of any number of teams (including none at all), and there's no limit on the number of teams, so you can create as many teams as makes sense for your environment. So, for example, if you had teams for *Finance*, *Developers*, and *Managers*, then the head of the finance department would be a member of both the Finance and Managers teams. But an individual software developer would be a member of only the Developers team and no others. You might even have some users who are in all teams (like the CEO) and some who aren't members of any (like salespeople in this particular example).

To manage teams, you'll need to be logged in as an account with the admin role. Then select the "Teams" item from the configuration menu. Once there, you can create a new team by clicking the "New Team" button, as shown in Figure 5-8. (If you're creating your first team, you'll see this button in the center of the view; after you've created one team, it will remain where shown in Figure 5-8.)

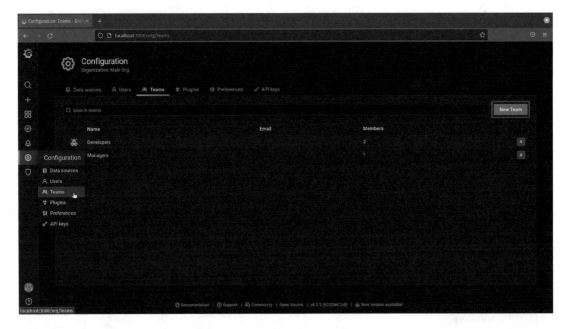

Figure 5-8. *The Teams configuration page in Grafana*

Clicking the "New Team" button will bring up a page prompting you for a team name and an optional email address as seen in Figure 5-9. The team name can be anything you like, as long as it doesn't already exist, so it's best to be descriptive but concise. Things like department names or regions are good examples of a team name.

The email address is used only to provide an avatar for the team and does not have to be filled in unless you want to customize this.

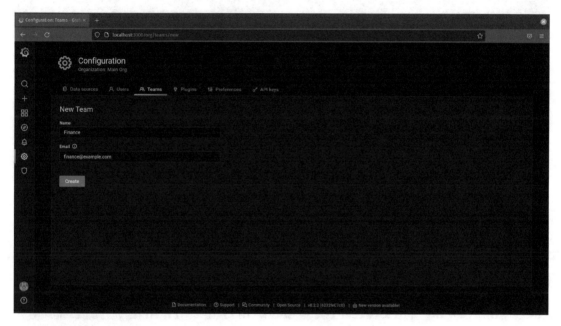

Figure 5-9. *Creating a new team*

Note   If you provide an email address, Grafana will look it up on the Gravatar service at `https://gravatar.com` and use the icon it finds there for the team. You can change icons by managing them on the Gravatar site.

To add users to a team, first click the team in the Teams configuration page. You can then click the "Add member" button to add users to the team. You can then either select users from the dropdown or begin typing a name to find the Grafana user you want to add to the team. Figure 5-10 shows an example of adding a user to a team.

Figure 5-10. *Adding a new member to a team*

Members can be removed from teams by clicking the red X button next to their name in the team page. Note that removing a user from a team does not remove them from Grafana or from any other teams.

Teams have a few settings associated with them as well, which you can see by clicking the "Settings" tab on a team page, shown in Figure 5-11. Here, you can change the team name or email address as well as provide some defaults for members of the team including the color scheme Grafana is presented in, their default time zone, and what dashboard loads when they first log in to Grafana.

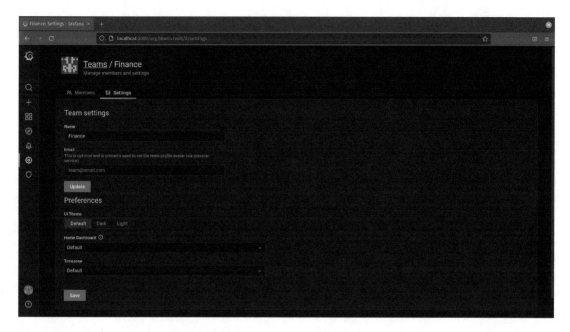

Figure 5-11. *A team settings page*

Finally, to delete a team, just click the red X next to the team name on the Teams configuration page. Deleting a team will remove its settings and configuration, but will not delete users or dashboards from Grafana.

Managing Permissions

Assigning users to roles in Grafana provides a very broad way of setting permissions. But frequently you'll want to be more nuanced. Some users might need to be able to edit one or two dashboards but not the rest. Or you might want to make certain dashboards available or hidden based on the user's role in your organization. Fortunately, Grafana provides a way to do all of this!

There are several ways that default permissions and dashboard permissions overlap, so let's walk through this with an example. Say you've created a dashboard with sensitive financial information that should not be seen by everyone in your environment. By default, all dashboards are viewable to anyone with at least the viewer role, which is everyone who can log in to Grafana. So we'll want to change those permissions to do a few things:

- Remove the default view permissions from the dashboard

- Add in view permissions for the specific users who need to see the data

- Add in edit permissions for the users who need to make changes to the dashboard

There's a lot to do, so let's get started!

To change the permissions on a dashboard, start by loading that dashboard in your browser. You'll need to have admin access to this dashboard, meaning you'll need to be logged in as an account that has the admin role, or you'll need to have already been granted admin permissions for this specific dashboard. If you're logged in as your admin user, you'll have this permission already.

Note   You may have noticed when saving your dashboards that they can be organized into folders in Grafana. This means that instead of having to manage permissions for individual dashboards, you can manage them on whole folders at a time. We'll look more at organizing dashboards and using folders in the next chapter, but for now just know that all the permissions settings we're using for one dashboard can be applied in exactly the same way to a whole folder.

Once you've loaded the dashboard, click the gear icon in the upper right of the screen to bring up the dashboard settings page. Figure 5-12 shows our financial dashboard with the settings icon highlighted.

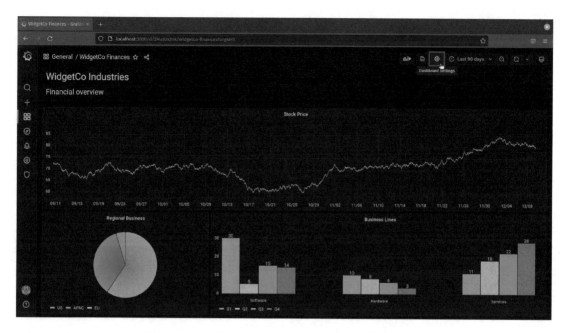

Figure 5-12. *A dashboard with the settings icon highlighted*

You'll see a number of possible settings here. We'll look at some of these settings in more depth in future chapters, but for now we're interested in the "Permissions" section, which is one of the links on the left. Clicking this will bring up a list of permissions associated with the dashboard. Figure 5-13 shows this list as it's set for all dashboards by default, with admin, editor, and viewer roles assigned their standard permissions.

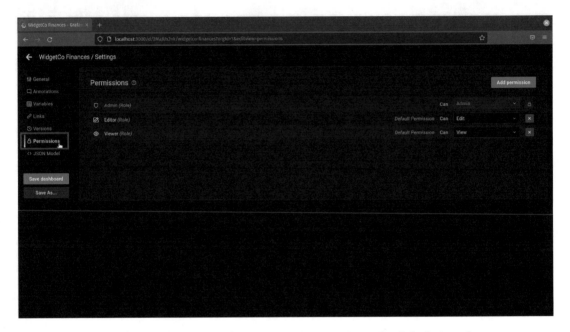

Figure 5-13. *The dashboard permissions settings page, highlighting the "Permissions" section link*

The first step we want to take is to remove these default permissions, as otherwise anyone would be able to view this sensitive data. To do this, click the red X button next to the default editor and viewer roles. (Note that you can't remove the admin role. An admin is always able to see and make changes to any dashboard.)

Once that's done, we can add in permissions for the users who should have access. You can assign permissions for both teams and individual users, both of which are managed by clicking the "Add permission" button.

For this dashboard, let's add two teams, Managers and Finance. Our managers need to have access to this sensitive data, but they don't need to make any changes. So we can add the view permission for this team by selecting the Managers team in the team dropdown and the View permission in the permission dropdown.

We'll also add our Finance team as editors, as they need to make changes to the dashboard from time to time. Figure 5-14 shows the Managers team already added and the settings for the Finance team ready to be added to this dashboard.

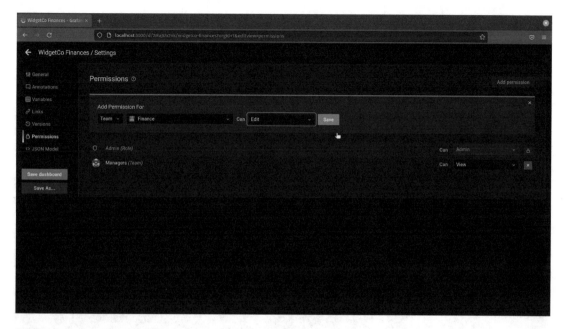

Figure 5-14. *Setting team permissions on a dashboard*

If you want, you can set permissions for individual users as well by changing the first dropdown from "Team" to "User." This works exactly the same as before, but instead of assigning permissions to a team, it assigns them to a specific user.

Tip   It's usually best to manage permissions with teams instead of with individual users. If someone joins or leaves your organization or transfers to a different role, you can update their permissions in one place instead of having to check every single dashboard in your environment. Unless there's a very good reason, try to avoid adding permissions for individuals.

If you've made any mistakes or want to change the permissions, you can do that from this screen exactly as you can for individual user roles: the dropdown changes the role assigned to a user or team for this specific dashboard, and the red X button removes the user or team role from this specific dashboard.

Now that the permissions have been applied, you can test them by logging in as different user accounts that are members of these teams. If a user is part of a team that has at least view permissions, they should see the dashboard normally. Other users won't see the dashboard at all, and trying to load it will show a permission denied message. Figure 5-15 shows a side-by-side view of a user with access to the dashboard and a user without the appropriate permissions trying to load it. Note that the URL is the same in each window; the only difference is how the permissions we set are being applied by Grafana.

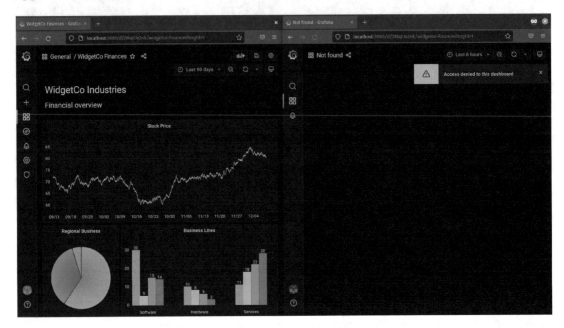

Figure 5-15. *A user with permission to view the dashboard on the left and a user without permission on the right*

Grafana Organizations

Organizations are another way of grouping and managing access to Grafana resources. A Grafana organization is a set of data sources, dashboards, and user permissions all grouped together. You can think of it like a separate Grafana instance that runs "inside" the first Grafana instance, sharing only the actual server and optionally user accounts.

Breaking this down a bit more, we can look at what is and what is not shared between organizations in the same Grafana environment:

Shared

- Physical and logical infrastructure (as organizations are part of a single running Grafana instance)

- User logins (though permissions are tracked separately)

Separate

- Data sources

- Dashboards

- Roles and permissions for user accounts (though accounts can be shared across organizations)

Organizations are rarely used. They're only really useful when you want to have completely separate environments but be able to use a single login to access all of them. One example would be if you are running Grafana and want to host dashboards for your customers. Each customer would have their own set of dashboards, their own data sources, and their own settings, and you wouldn't want one customer to be able to see another customer's data. But you as the owner of these need to be able to see all customers' data. In this case, you could have an organization for each customer where customers have access only to their own organization, but your account has access to all organizations.

Note that this doesn't give your account the ability to see all the different customer data at the same time! Organizations act like separate Grafana environments, so you need to switch between them when you want to see different customer dashboards.

Note   Organizations are one of the few areas where a Grafana environment that you deploy and manage yourself and Grafana Cloud differ. Grafana Cloud does not provide organizations; instead, it allows you to create and manage *stacks.* A stack is a set of hosted Grafana services including Grafana itself. If you want to have the functionality that organizations provide in Grafana Cloud, look into setting up more stacks there.

To manage organizations, you need to log in as an account with the server admin role. By default, this is only the "admin" user account. Even other accounts that are marked as having the admin role in Grafana do not have the server admin role.

Once logged in as a server admin, you'll see the Server Admin menu icon in the navigation bar on the left. Navigate to the "Orgs" item to access the Grafana organization settings page, as shown in Figure 5-16.

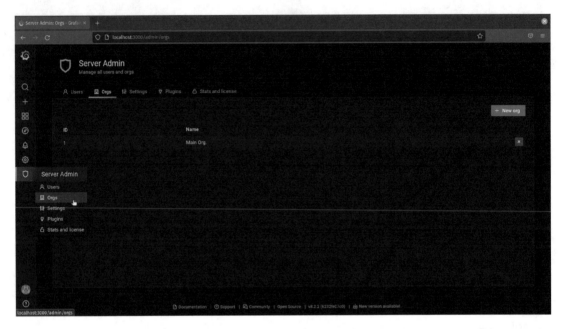

Figure 5-16. *The Server Admin menu showing the Grafana organizations settings page*

Note   You'll probably notice that there's an organization already there called "Main Org." Every Grafana instance actually has at least one organization, and this is the default one. Normally, it's not something you notice because unless there are multiple organizations configured, the options to switch between them are hidden. You'll only see these if you have access to at least two organizations.

To create a new organization, click the "New org" button. This will bring up a screen prompting you for a name for the new organization. Much like with teams, you can call an organization whatever you like as long as that name isn't already in use.

Once you create the new organization, you'll be immediately logged in to it so that you can start configuring it. If you look around at users, dashboards, and data sources, you'll notice that there's nothing there – it's like a fresh Grafana install. The new organization doesn't contain anything from the original organization and will need to be set up from scratch.

To switch between organizations, open the user menu near the bottom of the navigation bar (just above the help icon). You'll see an entry for "Switch organization" which will allow you to log in to any organization that your account has access to, as shown in Figure 5-17. Doing so will effectively log you out of the current organization and into the new one. (Note that any users that have access to only one organization will not see this menu; it only appears when it's actually useful.)

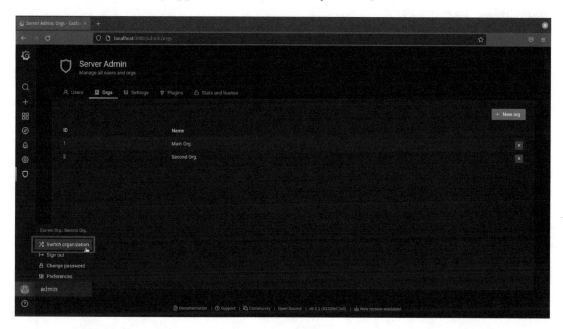

Figure 5-17. *The user menu with the "Switch organization" item highlighted*

To delete an organization, first switch to a different organization. (Grafana won't let you delete the organization that you're currently logged in to.) Then click the red X button to delete the organization.

Caution   Unlike deleting teams, deleting an organization **will delete everything** inside that organization. All dashboards, data source configurations, user roles, and permissions – everything will be gone. Be sure you really want to do this before you delete the organization!

Summary

In this chapter, we've looked at how to manage users in both Grafana directly and Grafana Cloud. You've seen how to manage roles and permissions, to set group memberships, and apply group and user permissions to dashboards and folders. You've also seen how to create and manage Grafana organizations for even more control over your Grafana resources.

In Part III (Chapters 6–11), we'll move beyond installing and configuring Grafana. We'll start to look at how to use principles of dashboard design to make appealing and useful data visualizations and how to create workflows to guide users to the level of detail that they need. We'll look at some more advanced data and dashboarding topics and finally learn how to create automated alerts based on the data that is accessible to Grafana.

PART III

Making Things Useful

So far, we've looked at how to deploy Grafana locally and how to consume it as a service through Grafana Cloud. We've also seen how to connect to other systems and run queries to pull data into Grafana for visualization. We've even seen how to share access to Grafana with other individuals or even publish it out to an entire organization.

Now we can take these building blocks and start constructing something truly useful!

In Part III (Chapters 6–11), we'll explore how to think about dashboard design, surfacing the most important data to our audience quickly and accurately while still giving them the ability to see more information when they need it. We'll then see how to connect multiple dashboards together, enabling users to move from a high-level view to specific details and even into other tools outside of Grafana. We'll also see how to build a reusable library of visualizations to make dashboard design faster, easier, and more accurate.

After this design overview, we'll look at how to combine data from multiple different sources in a meaningful way, even deriving new data from two sources. We'll then return to dashboards one more time, this time diving deep into templating, formatting data with dashboard variables, and even working with real-time streaming data.

With that thorough understanding of data and dashboards, we'll be ready to look at how to set up alerts in Grafana to notify users when issues arise. You'll then be able to build a full observability strategy, from data collection through visualization, issue notification, and remediation.

CHAPTER 6

Dashboard Design

There are two main goals to keep in mind when designing dashboards: they should be *useful* and *beautiful*.

Being *useful* seems obvious. A dashboard should contain relevant information that helps a person viewing it to understand the current state environment or system being monitored and (usually) what ways it's changing over time. We need to have the right data and know its history to know what's going on.

But just having the right data isn't enough. Being *beautiful* is just as important! By "beautiful" we mean that the data needs to be presented in a logical, consistent, and compelling fashion. It's easy to scan quickly. A beautiful dashboard gives indications of what is expected and what is unusual, shows what's most important and what can normally be ignored, and groups related information together.

Consider the two dashboards in Figures 6-1 and 6-2. Both contain the exact same information. Figure 6-1 presents the raw data as it's collected, while Figure 6-2 takes this data and applies principles of visual design.

© Ronald McCollam 2022
R. McCollam, *Getting Started with Grafana*, https://doi.org/10.1007/978-1-4842-8309-7_6

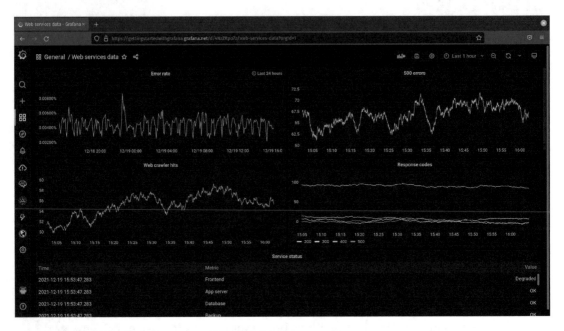

Figure 6-1. *A poorly designed dashboard*

Figure 6-2. *The same dashboard with design principles applied*

While the first dashboard shows all of the data, no real thought has gone into how viewers will use this data. It uses default, undifferentiated views of the information it exposes but doesn't give any additional help or context to understand it. You certainly can figure out what's happening here if you're willing to put in the effort, but it's going to take a lot of work.

The second dashboard shows the same data but with considerably more thought put into helping the viewer understand the data. It definitely takes more time and effort to build this sort of dashboard, and it's probably one that will see multiple iterations before it's fully polished. But that time and effort pay off every time that dashboard is used.

A well-designed dashboard takes all of this into account. All of the relevant data is presented, but the most important data is highlighted. Related information is grouped together to make drawing conclusions faster. Appropriate visualizations are used and wherever possible take advantages of affordances such as threshold markers or consistent color schemes to add additional context. Clear labels and sometimes even descriptions are available. And whitespace and headers are used to make it easy to scan.

Choosing Data

Fundamentally, dashboards exist to make data accessible. But if the data is wrong, incomplete, or even just presented misleadingly, it can lead people to draw incorrect conclusions. It's important to make sure that the data you are showing is correct before going any further.

It's often tempting to just grab the data that you have on hand and start layering visualizations on top of it. While there's a time and a place for this – usually in early explorations or to explore data for fast incident resolution – it's not the best way to build a permanent dashboard.

To start, think about what you want to show. You'll need to consider three things: your *audience*, the *data* you have access to, and the *context* that the data exists in.

Audience

Your audience is the most important consideration when choosing the data to present. If your dashboard doesn't suit its viewers' needs, then it's simply not a good dashboard. So think about what the people using your dashboard will need to know.

If your target audience is technical, say a Site Reliability Engineering (SRE) team, they will need to know things like infrastructure utilization, request latency, and error rates. You can assume that an SRE knows what the applications they support do and how they connect, so giving them the low-level information about multiple services and individual requests flowing through the system can give critical clues as to where to find and fix bottlenecks or errors.

On the other hand, if your audience is a business user or a user of your application, exposing infrastructure details is not likely to be helpful. These users care more about whether a service is functioning normally and how any outages are impacting their own products or services. The CEO of a company probably doesn't care about CPU utilization in a specific Kubernetes environment, but they do care about how many shopping carts are being abandoned due to slowness on their ecommerce site.

For any nontrivial environment or system, it's unlikely that you'll be able to address the needs of all audiences with a single view. While you can always pack in more charts or tables, this adds visual noise for the people who aren't interested in it and actually detracts from the usability of your dashboard. It's almost always best to identify a single user for your dashboard and keep them in mind when designing it. Don't try to cram everything into a single page; create a new dashboard for each audience.

Data

Once you've identified your audience, you can start thinking about the information that they want to see. Try not to fall into the trap of collecting all the data that you have and just formatting it. Instead, come up with a comprehensive list of the data that you need to be able to provide the viewer with a complete view of what's happening in their environment.

Tip   If you're not the target audience for this dashboard, do everything you can to work with the right people when identifying data. This can include bringing them into the design process directly, but at the very least be sure to interview some people and have them give you feedback as you design your dashboards. If you don't have this sort of input, you can put a lot of time and effort into building something that's not actually useful – nothing is worse than putting hours into building something only to see that it sits unused because it doesn't meet your audience's needs!

A good exercise here is to start with the end result that you want and work backward to the data that supports it. Going back to our previous examples, the SRE team wants to understand the error rate, utilization, and latency of the services they support. So in order to convey the status of this service, we know that we need to capture data like the number and type of errors from every service in the application, number of users hitting the service, and number of requests that are going through (both generated by end users and internal requests between services in the environment), and we need to have time spans covering all of those. We could keep digging deeper from there into interservice relationships (we probably care about different metrics on a database than on a part of the application logic) and might even need to capture data from the end users (such as web browser or connection speed) in order to draw good conclusions about where issues are occurring.

You'll often find by doing this that you don't have all of the data that you really need! This is the right time to start adding more instrumentation to your application or finding other sources of data to add in. In order to tell the whole story, you'll need all this supporting information. If you start off by just looking at what data you already have available, you'll often miss critical information that you didn't realize you needed.

It's also important to make sure that the data that you have is correct. It should go without saying that conclusions drawn from incorrect data will themselves be incorrect, so it's worth verifying up front that you really do have the data you think you do. Be sure also to double-check any information that's derived from multiple different sources. Even if your underlying data is correct, comparing data that's averaged per second with data that's averaged per minute can be tricky!

Context

Once you know who your audience is and have all the information you need to provide to them, it's time to think about how all of this fits into a larger picture. No person or system sits in total isolation, but you can't put the entire world into a single dashboard. So you need to consider how the data you're showing fits into the wider picture.

Understanding the context that your dashboard will be used in helps you synthesize what data you should display to your audience. If there are critical pieces of information that are not going to be displayed in your dashboard, it's important to make that obvious (and ideally provide ways for users to get that information). If there are thresholds or limits to your data, you should think not only about what the numbers are but why they exist and what the impact will be if they are exceeded. Keeping that information in mind as you build your dashboard will help you convey the most important parts as you build the visualization.

A great way to be sure that you provide all of the context is to write out what each part of your dashboard will convey in text rather than pictures and describe *why* you are including that information in the first place. Give a short, meaningful title to each separate idea that you want to convey, then describe what the data looks like and why it's important that you're including it.

For our SRE team dashboard, we might include something like

- **Title:** Error budget burn rate.

- **Data:** Average errors/minute over 60 minutes, number of errors allowed by contractual service-level objectives (SLOs). Error data comes directly from web service logs stored in Loki, and the SLO number is stored per customer in the customer PostgreSQL database.

- **Units:** Errors (counter, whole numbers).

- **Description:** A comparison of how many errors are occurring in our application stack (averaged over the past hour), compared to the number of errors we consider to be acceptable before exceeding our error budget.

- **Reasoning:** We provide service-level agreements (SLAs) to our customers that guarantee them a certain level of performance. This performance level means that we can't exceed a specific number of errors in an hour. If we do exceed that, we have broken this SLA and will need to refund the customer for the hour(s) in which we were not meeting the SLA.

Looking at this description, it's immediately clear that the error budget burn rate is very important – if we go above a threshold, we're going to have to pay penalties to our customers! So it's likely that this will go near the top of our dashboard and will need to be presented in a simple, direct way that calls attention to any negative trends in the data. We also know exactly what data we need to display it and where that data exists. It mentions a customer database, so when we go to build the visual representation of this data for our dashboard, it might be useful to indicate that or even to provide a link to the database itself.

Finally, this exercise means that we understand the context in which this data is used. It's impactful to the SRE team but also likely to sales and customer support, so when using it in a dashboard we'll want to consider cross-linking between dashboards to expose more detail faster, which we'll explore in the next chapter.

Having this detailed description ready to go before building the dashboard lets you both think clearly about your design and get a lot of the mundane tasks like coming up with a coherent naming scheme for your panels out of the way early.

Composing a Visualization

Once you've identified your audience and collected the data that you need, you can start putting things into panels for visualization.

Dashboards should provide fast access to information. Your goal should be to present a view of the data that can be understood at a glance, or at least without too much analysis. In order to do that, your visualizations must be clear and concise.

Most importantly, each visualization should convey a single idea in simple visual form. Don't try to mix multiple unrelated items in a single graph. If you can't clearly express what you're showing in a few words, it's likely that you're trying to do too much in a single visualization. Try splitting it up into two or more panels. Even if multiple data sets are related, try to keep them as simple as possible. It's almost always better to have multiple graphs next to each other than to cram everything into one view.

If you're tracking the number of widgets produced per hour and the number of defective widgets over the same time period, these should be two separate charts. Even though the time axis is the same for both data sets, combining them into a single graph makes it hard to quickly understand either. Figure 6-3 shows a combined graph of error percentage and production on the left, with the same data split on the right. In this example, a dashboard option to show time alignments between panels has been enabled to further show how data can be correlated across graphs.

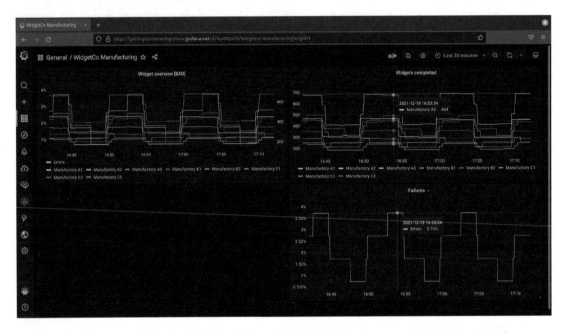

Figure 6-3. *A confusing combined graph (left) and two separate but related graphs (right)*

The exception to this is when the relationship itself is important information. If increasing the speed of widget production leads to a higher rate of widget defects, that's important to know. But it's hard to see from simply overlaying the two data sets. Instead, consider the widget failure rate as a function of widgets produced as its own time series derived from the original data. (We'll look at how to derive a new series from existing data in Chapters 8 and 9.) Overlaying this derived series on one of the original series can show meaningful information clearly. Figure 6-4 illustrates this, using simple lines to show manufacturing data from multiple machines and a gradient under a line to indicate error rates. Notice that as the error rates increase, the gradient becomes more opaque, giving a clear indication of the relationship between the series.

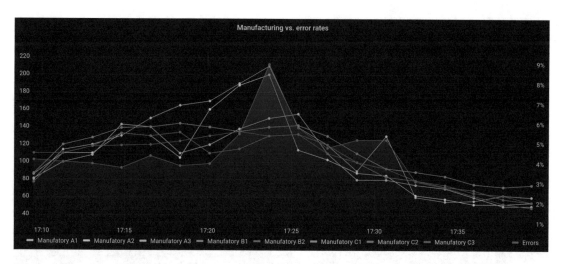

Figure 6-4. *A gradient used to indicate a series that is related to, but not part of, the core data set*

This example also shows how you can use visual indicators to convey context. By using shading, we can draw attention to the differences between two time series, effectively showing that the rate of change between the two itself is moving over time. Each panel type has its own visualization options, so be sure to explore these settings when creating your dashboards.

Be sure to be consistent in your use of visualization types and options. If you're showing error rates for two different services over the same time period, use the same visualization type for both. Keep colors consistent as well: if your US environment is green and your EU environment is blue on one graph, keep those colors the same on other graphs. If you use yellow as a warning threshold and red as a danger indicator for CPU utilization, use the same colors for disk capacity available. This will enable your viewers to quickly compare data without having to check the graph legend each time.

Finally, be sure to set units on your visualizations where appropriate. Grafana has a huge list of built-in unit types ranging from geometry and physics (angle, area, distance, acceleration, mass, temperature, and so on) to computing (operations per second, bytes, hashes per second, and so on) to percentages and even logical values like on/off or true/false. Taking a bit of time to select the appropriate unit for your visualization means that Grafana can automatically scale it to something that humans can understand. It's far easier to understand a graph showing units like megabytes or gigabytes instead of thousands or millions of bytes. Figure 6-5 shows the same data in the same visualization, but with the proper unit set on the panel on the right. Notice how much easier it is to understand the data when looking at the labels on the Y axis. As a bonus, Grafana will scale these labels appropriately to kiB or GiB as the data shrinks or grows.

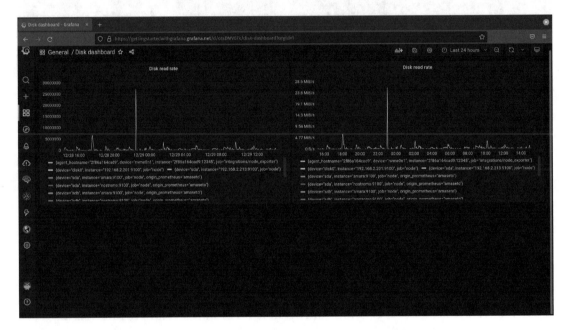

Figure 6-5. *Setting units appropriately makes panels far more useful*

Surfacing the Most Important Data First

An axiom of dashboard creation is that the most important data should be presented in the most prominent way. This applies at all levels of dashboard design.

In the dashboard as a whole, the most important panels should be at the top of the dashboard. This ensures that they will always be seen no matter the size of the viewer's browser. These are often the largest panels as well, and they should show a minimum of detail so that the most critical information is conveyed. Additional details and context can be provided by smaller panels below these top panels, giving the viewer the ability to dig into these details as needed. Use the "10-foot" or "3-meter" rule: if you were to put your dashboard on a display on the wall, the most critical information should be visible and understandable from about this far away.

Each individual panel should follow this pattern as well. Even though the data is the "interesting" part of your panel, the panel title is the headline. Be sure that your titles provide information about what the viewer is looking at and let them put it into context with the rest of the dashboard. Give relevant detail in your titles without going overboard: "Widgets produced per hour" is clearer at a glance than "Production," but "Sum of all widget production across each assembly line grouped by hour" is too long to take in quickly.

Consistency is key across your dashboard. Panels of equal importance should be of equal size and position. Panel titles should use consistent terminology. Keeping things the same across panels will help your viewers apply an understanding of one part of your dashboard to the rest, and they'll be naturally drawn to the most important data.

Adding More Context

Though the goal of a dashboard is to present information that can be understood immediately, there are times when additional details are useful. Additional context might need to be added to describe the data or to call attention to particular details. Even just a note informing the viewer where they can get more information or how to contact someone in case of an issue can be helpful.

The first and most important piece of context that you can add is a panel title. As mentioned earlier, this should clearly state to the viewer what they are seeing, but needs to be kept short enough to take in at a glance. For more complex visualizations, such as comparisons between multiple related data sets with different units of measure, it can be difficult to explain this in the few words allowed by a title.

Additional context for graphs can be added via the panel description field, as shown in Figure 6-6. This field can be used to explain details of your data visualization that might not be immediately apparent, provide additional instructions for interpreting or using the information it provides, who to contact about issues, or anything else that might be relevant. It also supports a limited set of Markdown features, so you can do things like emphasize text (here shown using underscores) and provide links (the brackets and parentheses used at the end of the text).

Figure 6-6. *Creating a description for a panel using Markdown syntax*

Once you've applied this description to your dashboard, you'll notice a small information icon in the upper-left corner of the panel. This will show your description when the viewer's mouse cursor hovers over the icon, as shown in Figure 6-7. Note that the Markdown syntax is applied, and any Markdown formatted URLs or raw URLs are turned into links that the user can click.

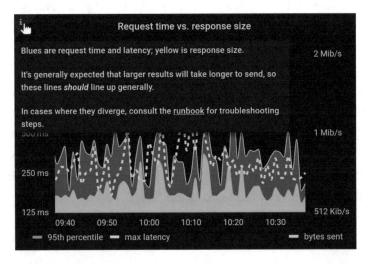

Figure 6-7. *A panel description with Markdown formatting and links*

Note   You might be wondering how to set options like line styles for a single data series or how to use multiple Y axes. This is done using *dashboard overrides*, which we'll look at in Chapter 9.

One other way to add context to your visualizations is by using text panels. Text panels allow you to do the same sorts of things that you can do with the panel description, but since they are actual panels, you have full control over their size and position. They're also displayed at all times, unlike panel descriptions which have to be revealed with the mouse cursor.

Text panels let you provide as much description or information as you like and can be very useful where you have to convey critical information about a visualization that should not be missed, or where you want to provide lists of links to other dashboards or external tools.

But overuse of text panels can also clutter a dashboard and obscure the data on the dashboard. Be very careful not to use so many text panels that you distract from the actual information that you are providing!

Exposing Detail with Rows

So far, whenever we've added any item to a dashboard, we've always picked the option to create a panel. But you probably noticed the option to add a *row* as well.

Rows provide two functions in Grafana. When used with *variables* (which we'll cover in Chapter 9) they can repeat a set of panels for multiple different sources or series of data. This lets you create a dashboard that will adapt to the data you feed into it without you having to create the same set of panels over and over.

The other function of rows is to provide a way to organize and hide or display data on a dashboard. Rows act as containers for panels and can be shown or hidden as a group. This makes it easy to create a dashboard with supporting details that are normally not displayed, but can be accessed with a single click when needed.

Let's take a look at how rows work. Figure 6-8 shows a dashboard with a lot of detailed tabular data supporting a few more interesting charts at the top. For this example, we'll assume that the tabular data really is relevant and accessed somewhat frequently and that it can't be displayed in a more viewer-friendly format. But it's not something that we can really understand at a glance, and when looking at the more important charts at the top, this just creates visual clutter.

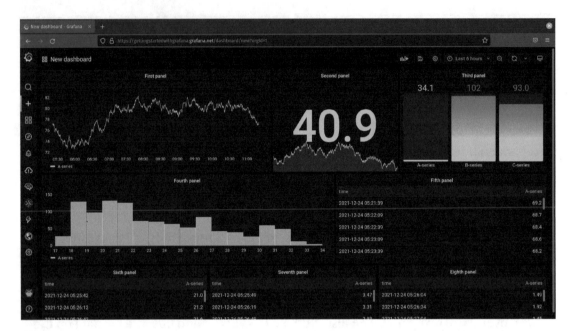

Figure 6-8. *A dashboard with noisy supporting details*

We can use a row here to hide the less important data but still have it be available when needed. Figure 6-9 shows what the same dashboard looks like after adding a row to hide this visual clutter by default.

Figure 6-9. *A dashboard with the noisy supporting details hidden but still available*

The details are still there, but they are hidden inside of the "Supporting details" row and can be brought back with a single click.

Rows work slightly differently than normal panels in Grafana, so we'll look more closely at how to use them effectively.

Adding Rows

To add a row, first start by clicking the "add panel" button, just as you would to add a normal panel. But when presented with the panel selection menu, choose the "Add a new row" option instead, as shown in Figure 6-10.

Figure 6-10. *The add panel selection menu with the new row option highlighted*

After selecting the new row option, you'll see a small space added to the top of the dashboard with the text "Row title." This is your new row, and by default it will add all of the panels beneath it to the row. This means that all of the panels you previously created on this dashboard are automatically added to the row.

To see what this means, we'll close the row which will hide the panels inside it. To do this, click the small indicator to the left of the row title. When open, it will look like >, and when closed, it will appear as a **v**. Figure 6-11 shows two side-by-side views of a dashboard with all panels inside of a row. On the left, the row is in its default open state, and on the right it is closed, hiding the panels.

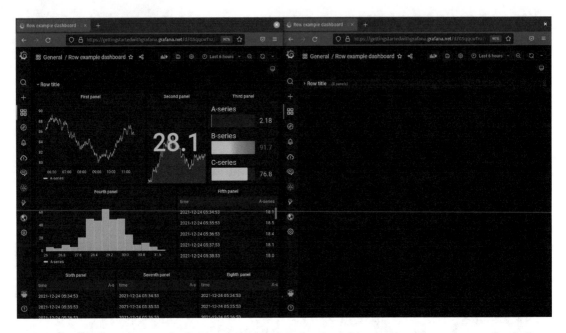

Figure 6-11. *Two views of the same dashboard with panels in a row; on the left, the row is open; on the right, the row is closed and its content hidden*

Configuring and Deleting Rows

Unlike panels, rows don't have a lot of options available. You can really only change their names and configure what variables they use, which we'll look at in Chapter 9.

To configure a row, hover over its title until you see the row options icon (which looks like a small gear). Click this icon and you'll see the row options menu, as shown in Figure 6-12.

Figure 6-12. *The row options menu*

Changing the title field will update the row title on your dashboard. The other option works with variables, which we'll explore later.

To delete a row, hover over the row title and click the delete icon (which looks like a trash can). You'll be prompted as to whether you want to delete all of the panels inside the row or just the row itself. If you delete everything, all panels in the row will be gone from your dashboard, but otherwise the row will be removed and the panels will return to the dashboard outside of any rows.

Moving Rows

Rows are a unique object in Grafana as they can't be moved in their default open state. In order to move a row, you first need to close it. Once the row is closed, you'll notice a series of dots on the right side of the row, as shown in Figure 6-13. This is the row handle, and dragging it will let you move the row up and down the dashboard.

Figure 6-13. *A closed row with the row handle highlighted*

Note   You can only move rows vertically on a dashboard. They can't be resized or moved horizontally. They fill up an entire row of the dashboard, just as their name implies.

Moving Panels In and Out of Rows

To move a panel into a row, first make sure the row is open. Then just grab the panel title and drag it below the row title. Any panel that is positioned under a row title will automatically become a part of that row.

To move a panel out of a row, click the panel title and drag the panel above the row title. Panels that are placed above rows will automatically be removed from the row.

Panels that are inside a row can be resized or arranged by dragging them around, just the same as if they were on a dashboard without rows. Any arrangement you make inside of a row will be saved, so that when the row is opened you'll see your panels in the layout you selected.

Tip   When moving panels around on a dashboard with rows, it can be useful to make sure all the rows are closed before moving panels. This will keep you from accidentally putting a panel inside of a row. Be sure to open any rows you want to be open by default before saving the dashboard – the open/closed state of rows is saved with the dashboard settings.

Using Multiple Rows

You can add as many rows as you like to your dashboard. Rows can't contain other rows, so you don't need to worry about losing panels inside of nested rows.

Moving panels between multiple rows works exactly the same as moving panels with a single row. If you drag a panel above a row but there's another row open above it, that panel will become part of the higher row. So it's easiest to make sure all rows are closed except the row or rows you're actively working with when moving panels around.

Tip   Use rows sparingly, as too many can add as much visual clutter as the panels you're hiding. If you find yourself adding a large number of rows to a dashboard, you should consider splitting it up into multiple dashboards. Don't try to do too much in one place, or you'll overwhelm your viewers!

Organizing Dashboards

One final consideration when designing dashboards is how your viewers will find them. As your use of Grafana grows, so will the number of dashboards (and probably the number of people creating them). Once you get more than a few dashboards, it can be tough to sort out which one you're looking for in a hurry and even harder to notice when something new has been added.

Fortunately, Grafana provides ways to organize and add metadata to your dashboards. These come in the form of *folders* and *tags*. You can use either or both of these to make your dashboards easier to discover. While they're both optional, they can help your viewers a lot. So think about how to organize your dashboards early and be sure to keep up with your system!

Folders

Folders in Grafana are similar to the folders in the filesystem of your computer, or of the paper files they're named after. Each dashboard exists in a folder, which by default is the "General" folder. Dashboards can be moved between folders, but each dashboard can only be in a single folder at a time.

Caution   You can make multiple copies of a dashboard and put each copy in a different folder, but these copies can quickly get out of sync. Changes made to one copy will not be reflected on another unless you manually make those same changes. If you find yourself wanting to have the same dashboard in multiple places, consider using tags instead.

Dashboard folders can be viewed and managed by using the dashboard browser, as shown in Figure 6-14. To access the dashboard browser, select the dashboards menu item from the navigation bar on the left and choose "Browse."

Figure 6-14.   *The dashboard browser and the dashboards menu item with the browse option highlighted*

To add a folder, use the "New folder" button. You can add as many folders as you like, but bear in mind that having too many folders to search through is just as bad as having too many dashboards in a single folder, so think about an organizational system that will work efficiently for you.

To move dashboards from one folder to another, click the checkboxes beside the dashboards you want to move. Once you've selected at least one dashboard, a move item will appear. Figure 6-15 shows the move button and the menu that is presented after clicking it.

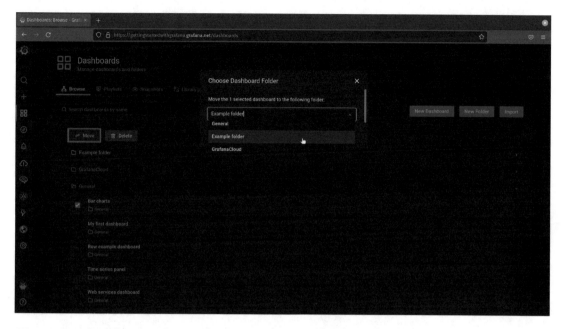

Figure 6-15. *The dashboard move button highlighted with the move menu displayed*

Note that folders are a single level in Grafana. You can't put folders inside of other folders, so keep that in mind when planning your organizational structure.

Tags

Tags are short text labels that can be applied to dashboards. Unlike folders, a dashboard can have as many tags attached to it as you like. Tags can also be attached to as many dashboards as makes sense.

To manage tags on a dashboard, start by opening that dashboard and then going to the dashboard settings menu via the gear at the top right of the dashboard page.

Once inside the dashboard settings page, you can add tags by typing an entry in the tags section as shown in Figure 6-16. Once you're finished with a tag name, hit enter to add it to the dashboard. You can add as many tags as you like here.

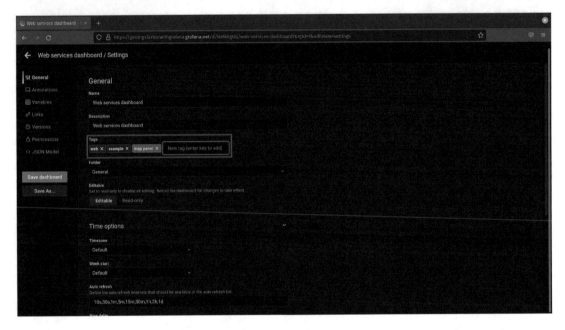

Figure 6-16. *Managing tags on a dashboard*

To remove a tag, click the small "x" next to the tag name.

Be sure to save the dashboard after adding or removing tags, or they will disappear when you leave the page.

Finding Dashboards

Once you've set up folders and tags, they can both be used to help you find relevant dashboards quickly. We've already seen how folders can be used in the dashboard browser. But folders and tags are also available in the dashboard search interface.

To open dashboard search, click the magnifying glass icon in the navigation bar on the left. This will open the search interface as shown in Figure 6-17. Putting text into the search box at the top will search both folder names and dashboard names. Using the tag menu on the right will narrow the search to only dashboards with a specific tag. By using both of these together, you can search a large number of dashboards in a very short amount of time.

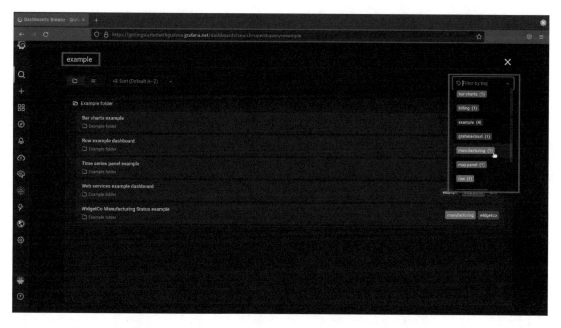

Figure 6-17.   *The dashboard search interface with the search bar (top) and the tag list (right) highlighted*

Summary

In this chapter, we looked at the fundamentals of good dashboard design in order to create dashboards that are both useful and beautiful. We explored using rows to organize panels on a dashboard and folders and tags to organize dashboards in Grafana.

In the next chapter, we'll extend these concepts into building workflows for dashboard viewers to follow to help them get the most out of the information you're presenting.

CHAPTER 7

Workflow

A beautiful dashboard is a great start! But you'll almost certainly have more information to show than can fit in a single dashboard. On top of that, any sizable environment will have different audiences to cater to – an executive probably cares more about the overall status of the environment and what customer impact an outage is causing, while an engineer wants more detail about the specific component they work on. In this chapter, we'll see some ways to satisfy everyone and surface relevant information by giving users a natural flow from high-level dashboards to more detailed ones.

Overviews vs. Details

So far, most of the dashboard design we've looked at has targeted overview dashboards. Giving the most important snippets of information first and hiding the details is great for that sort of dashboard, but there are times when you need to go deeper and make use of every bit of data that you have available.

It's certainly possible to build dashboards that serve both purposes. In the previous chapter, we looked at using dashboard rows to hide and show data. This is a pattern that is useful up to a point, but gets unwieldy quickly.

Consider a dashboard that shows the status of a large company's entire ecommerce environment. This dashboard needs to be useful to everyone up to and including the CEO, so it should show the most important data right up top. In this case, that's probably information like whether the site is up and reachable, how many sales have occurred in the last day or week, the number of concurrent users online right now, and so forth.

But people other than managers need to use this dashboard too, and that's where things can get tricky. There are a bunch of engineering teams that care about the details that roll up to these numbers. But while the web design team cares a lot about whether the new site designs are increasing sales over time, the infrastructure team is more worried about CPU and memory utilization on the underlying servers. The database

© Ronald McCollam 2022
R. McCollam, *Getting Started with Grafana*, https://doi.org/10.1007/978-1-4842-8309-7_7

administrators, on the other hand, really want to see which queries are slowing down transactions and where they're coming from. And that's not even yet thinking about customer engagement (who cares about satisfaction survey results and how they correlate with cart latency), the shipping department (who like the CEO cares about sales numbers, but *really* needs to see what bulky items are being bought fastest and what the stock numbers are), fraud prevention (who want to look for unusual spikes in sales in specific regions), and any number of other consumers of this data.

This list of consumers of data is incomplete, but it's already overwhelming! Trying to put all of these different uses into a single dashboard would be equally overwhelming. By the time you crammed everything in, the dashboard would be too big to be useful to anyone. But you also don't want to have to maintain dozens of separate dashboards for each use case.

The Grafana approach to this problem is to branch out and diversify. You want to build a small number of dashboards – maybe even just one – for people to use as a starting point, but then give them an easy way to move into other more specialized dashboards. Having a clear path from the high-level overview to more and more detailed data provides users with a *workflow*, a way to be guided to the information they need with a minimum of effort. Grafana provides some useful tools to build these workflows, which we'll look at later.

There's one key idea to keep in mind when building workflows: good dashboard design principles don't change. Even though you're making a dashboard to show more detailed information, there are still some bits that are more important or interesting than others. As you build levels of dashboards that are increasingly detailed, be sure to stay consistent in your design and keep your dashboards useful and beautiful. A little extra effort up front will make your life and the lives of your dashboard's users significantly easier in the future.

Links

Links are one of the primary tools for building a workflow in Grafana. These work just like links on a web page, giving you a place to click to be taken to more information. In Grafana, links can point to other Grafana dashboards, to websites, or even to other monitoring tools. Links can also optionally pass along data supplied by panels on the dashboard like time ranges, values, or field names. Adding links to panels and data is a fast way to build up powerful workflows.

There are two types of links in Grafana: *panel links* and *data links*. They each provide a clickable link, but the positioning and behavior of those links differs between the two.

Note   In this section, we'll look at how to embed information from a panel into a URL using *variables*. For now, we'll start with how to put information into variables, but we'll look at how to retrieve the data and use these values in Chapter 9.

Panel Links

Panel links are the simplest link type in Grafana. As the name implies, they provide a way to add a link to a website or another dashboard from a dashboard panel. Like panel descriptions, panel links appear in the upper-left corner of a panel once they've been added. Figure 7-1 shows a panel with two panel links configured.

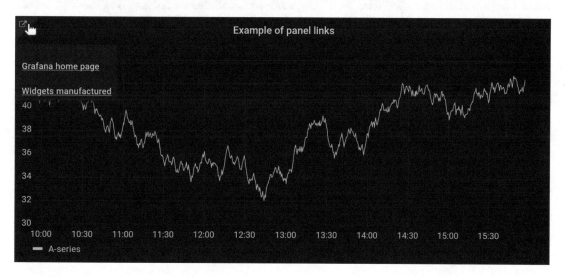

Figure 7-1. *A panel with two panel links*

To create or manage panel links, start by editing a panel. In the panel options, scroll down until you see the "Panel links" section, shown in Figure 7-2. In this example, there's one panel link already pointing to the Grafana home page.

Figure 7-2. *Editing a panel with the panel links section highlighted*

To add a panel link, click the "Add link" button. This will open a configuration screen for the new link, as shown in Figure 7-3. There are three items to fill in here. The title field will be the link title, that is, what the user will click to activate the link. The URL is the target of the link. For links to other Grafana dashboards, you can either put the full URL or copy only the part after your domain name. In this example, I've linked to another dashboard in my environment but left out the "*https://gettingstartedwithgrafana.grafana.net*" part, but as it's an internal link, this will still work. Finally, you can choose whether the link should open in a new tab or to reuse the existing dashboard window.

Figure 7-3. *Adding a panel link to another dashboard*

The URL field has an additional feature that isn't immediately obvious. You can pull some information from the dashboard itself in the form of *variables*. Variables contain information that the panel as a whole "knows" and can pass along to other dashboards or external tools. These variables will be replaced with an actual value when the link is clicked. For example, using the ${__url_time_range} variable in a URL will replace that string with the time range selected in the current dashboard. Adding this to a link to a Grafana dashboard will automatically set the same time range when opening the new one. This can save time and effort when moving to more specific dashboards when troubleshooting an issue, for example.

In the case of panel links, there are a limited number of variables available. These are documented at *https://grafana.com/docs/grafana/latest/linking/data-link-variables/*, though for panel links, only the set of time range variables are available.

You can also see the available variables by pressing control+space (or command+space on a Mac) when editing the URL field, as shown in Figure 7-4.

Figure 7-4. *Adding a panel link from the list of available variables*

Data Links

Adding a simple link to a panel is useful, but you'll often want more than that. If you're investigating an issue that's affecting one instance of an application with many normal nodes, for example, you'll probably want to focus your exploration on just the troubled server. Being able to pass things like names or data values into other dashboards or external tools would let you stay focused on the specific problem you're solving.

This is where data links come into play. Data links are similar to panel links in that they can direct the viewer to other dashboards or external tools. But instead of linking from a panel, they let you link directly from the data that is being displayed by clicking on the data shown inside the panel itself. Figure 7-5 shows an example of a data link on a time series panel, highlighting two pieces of information that are being passed in the link to another dashboard. In this case, an abnormally high amount of activity was seen on a specific disk in a specific server. Passing that data to another dashboard (the "Disk dashboard" option under the mouse cursor) can help build the workflow of moving from the general case (all disks in all servers) to the specific area of investigation (a potentially faulty disk).

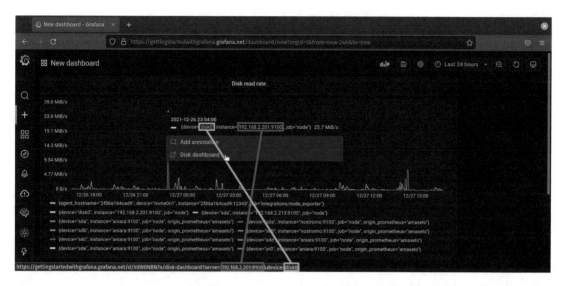

Figure 7-5. *A data link passing two pieces of information to another dashboard*

To create or manage data links, start by editing a panel type that supports data links. You'll need to find the data links section in the panel options on the right, as shown in Figure 7-6. In this example, two data links have been added. One points to another Grafana dashboard and the other to a site on the corporate intranet.

Figure 7-6. *Editing a panel with the data links section highlighted*

Tip While most of the default visualizations support data links, not all do. And when using panel types from the community, this functionality relies on the developer having enabled the functionality and written code to handle data links. If you don't see the data links section, try using another visualization type.

To add a data link, click the "Add link" button. This will open a configuration screen for the new link, as shown in Figure 7-7. Exactly as with panel links, you'll have three options to set. The title field will be the link title, that is, what the user will click to activate the link. The URL is the target of the link. For links to other Grafana dashboards, you can either put the full URL or copy only the part after your domain name, called a *relative link*. Finally, you can choose whether the link should open in a new tab or to reuse the existing dashboard window.

Figure 7-7. *Adding a data link. Data links provide far more variable options than panel links*

Like with panel links, you can show the list of available variables by pressing control+space (or command+space on a Mac). But you'll notice that the list is much longer here. This is because the user is not interacting with the panel as a whole but with a specific data point in that panel.

The information that can be sent includes things like the value that is selected, any metadata attached to a point (such as "environment," "server" or "instance," "application," or any other metadata that you have available), and even the name of the series itself (useful when you're showing multiple different sets of data on one graph). There's a lot to explore here, so you will definitely want to refer to the documentation on these variables at *https://grafana.com/docs/grafana/latest/linking/data-link-variables/*.

The Explore View

So far, any time we've looked at data, it's been in the context of adding panels to a dashboard. This is great when you want to refer back to a visualization that you've built again and again, but is overkill if you're just experimenting with data or want to answer a quick one-off question.

The Grafana *explore view* lets you run queries against a data source without the overhead of defining and saving panels to a dashboard, as shown in Figure 7-8. While it has fewer features than a full panel, it's faster to work with and gives an uncluttered view into the data that you're using. You'll also note that it shows both a graphical representation of the data and a tabular view, letting you quickly see specific data points as well as the larger trend.

Figure 7-8. *The Grafana explore view showing the results of a query as a line graph and as a data table*

The explore view can be a powerful part of a troubleshooting workflow. Frequently, you'll start by either directly noticing something amiss on a dashboard or being brought to a dashboard via an alert. As we've seen already in this section, a well-designed dashboard will give you the most important information and provide context about what's happening. But you'll often still need to inspect and query more deeply – asking questions of your data to truly understand what's happening. Moving from a dashboard to the explore view gives you a laboratory to run experimental queries.

Getting into the Explore View

The explore view can fit into a workflow in two ways: directly or from a panel.

The most obvious way to enter the explore view is to click the explore icon on the navigation bar on the left. The explore icon looks like a compass, as shown in Figure 7-9.

Figure 7-9. *The explore icon*

Accessing the explore view via the icon gives you a blank slate. No data will initially be loaded, and you can select a data source and begin running queries. This is most useful when you already know exactly what you're looking for and know the query or data to look up off the top of your head.

But you can also enter the explore view directly from a panel on a dashboard by using the panel menu, as shown in Figure 7-10. Opening the panel menu and clicking "Explore" will take you to the explore view, but will carry over everything from the panel you started from. This means that the query (or queries) displayed in the panel will be run in the explore view, and the time range will be exactly the same as it was on the dashboard you started from.

Figure 7-10. *Opening the explore view from a panel menu*

Tip For fast access to the explore view, you can use a keyboard shortcut. Hover over a panel on a dashboard with your mouse and hit "x" on your keyboard. This will immediately open the explore view with that panel's queries loaded.

Using the explore view in this way makes it a natural extension of a dashboard. You can start in a dashboard that contains a lot of information from various sources, but very quickly narrow (or broaden!) your focus to include exactly what you need to see.

Using the Explore View

The explore view is designed to make working with queries as fast as possible. It also exposes a few features that aren't found elsewhere in Grafana.

Running Queries

At its most basic, using explore is similar to building a panel on a dashboard. You select a data source at the top, then add one or more queries into the query box provided. As you're working with queries, especially complex ones, it can be helpful to refresh the data that you're seeing from time to time to make sure it's correct. To do this, you can click the "Run query" button in the upper right. This will execute the query as you have entered it but not make any other changes.

If you want to continually update the data (to watch ongoing trends or patterns), you can click the arrow next to the "Run query" button and select a time range. The data will be updated automatically at the frequency that you select.

Queries will also be logged in your query history, accessible through the "Query history" button immediately underneath the query itself. You can scroll back through previous queries that you've run rather than building them from scratch every time. If there are queries that you find yourself running frequently, you can click the star icon to save them, and they'll be available in your starred queries for quick access. Figure 7-11 shows a starred query being run in the explore view.

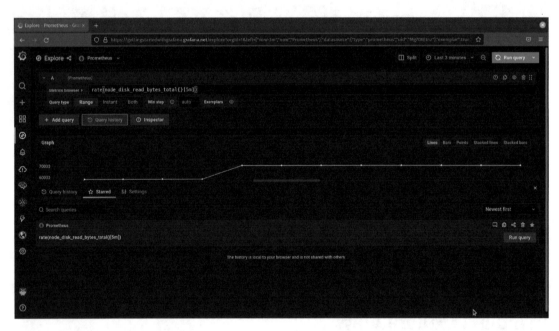

Figure 7-11. *Using starred queries in the explore view*

Updating Dashboards

If you started with a panel on a dashboard and entered the explore view from there, an additional option will be available. You'll have the ability to update the original panel with the query that you have altered in the explore view. Figure 7-12 shows the back menu that appears when entering the explore view from a panel and highlights the option to save your changes back to a panel.

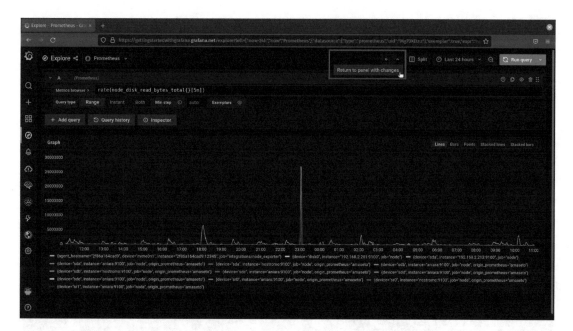

Figure 7-12. *Entering the explore view from an existing panel gives you the option to save your changes back to the original panel*

If you have made changes to the query that should be updated in the original dashboard panel, this is an easy way to make the necessary changes after you're finished revising the query.

Caution   Selecting this item will not automatically save the updated query in the original panel. You still need to hit save on the dashboard once you've returned! If you navigate away from the dashboard without saving, your changes will be lost.

Comparing Data with the Split View

The explore view also lets you compare and contrast multiple queries side by side, making rapid changes to each. To open the split view, click the "Split" button in the upper right of the explore view.

Once the view has been split, you can run two sets of queries side by side independently. Each side can have its own data source and time range defined. Grafana provides functionality to synchronize these views which can be a powerful way to troubleshoot complex issues.

Figure 7-13 shows the split view with these features highlighted. At the top is a clock icon, which allows you to set a time range for each query. By default, these are independent. But by clicking the lock icon next to this, both panes of the explore view will be synchronized. Changing the time range on one side will immediately update the other. This also applies when dragging the mouse across one graph to select a subset of data – if you narrow the time range by dragging on one side, it will affect the other side as well whenever the clocks are synchronized.

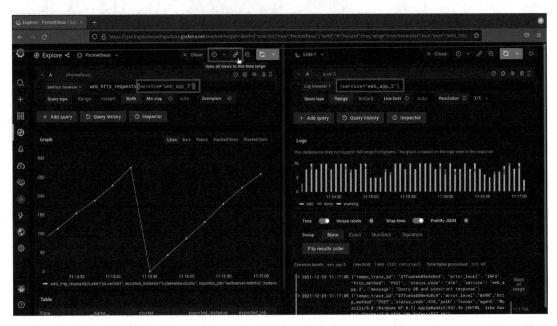

Figure 7-13. *A split explore view with the time range sync option and the automatic query sync functionality highlighted*

Grafana goes one step further though. It will also automatically synchronize queries wherever possible. For example, in Figure 7-12 we started with a metric query in Prometheus, looking at web response metrics from a specific web application. When splitting the explore view and selecting the Loki data source to look at logs, Grafana has automatically created a query to show logs for the same web application in the same time range.

Note   Not all data sources can be linked this way. It requires that the data sources share some common metadata and that queries with this metadata can be built automatically. This is automatically the case for Prometheus, Loki, and Tempo, but can be configured for many other data sources. If the data source exposes a "derived fields" section in the data source configuration page, this is where these links can be configured. Consult the plugin documentation for more information on configuring these fields.

Reusability with Library Panels

So far, our workflows have been focused on people viewing and using data directly. But it's equally important to consider the workflows of people creating new dashboards. Making it easy for people to find and reuse existing content means that they won't waste time and effort creating things that already exist. And providing a mechanism for people to get updates when changes are made to existing visualizations means that everyone stays in sync.

The Grafana approach to this is to use *library panels*. A library panel is a panel that has been created and managed centrally. It's easily found by searching or filtering when adding new panels to a dashboard and, most importantly, links to all copies of itself on a dashboard so that when the original library panel is updated, all copies of it can also be updated at the same time.

Library panels roll up both the query and the visualization of the data into one object. So any customizations or changes to the visualization (e.g., color schemes, thresholds, data transformations, etc.) will be applied to any panels that are added to a dashboard from the library.

Adding a Panel to the Library

To add a panel to the library, start by creating it like any other panel. You can either create a dashboard for just your library panels or use a panel from a dashboard that you've already created.

Once you've created or selected your panel, open the panel menu, navigate to the "More…" submenu, and select "Create library panel" as shown in Figure 7-14.

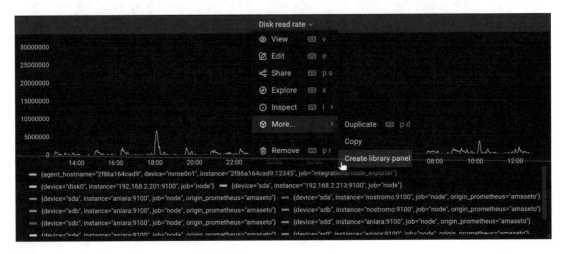

Figure 7-14. *Adding a panel to the library*

You'll now be given the option to name your panel and to select which folder it will be saved to. The name will default to the current panel name as it is on the dashboard, but it's a good idea to add a bit of additional context here. You might have more than one panel showing logged in users per hour if you have multiple applications, for example. Using a prefix "eCommerce – Users per hour" to differentiate it from "Intranet – Users per hour" can be helpful when searching for this panel later.

The folder that you select will be used to apply permissions to your library panel. If you add it to a folder that only admins can write to but everyone can view, you'll have a set of known good panels that can't be accidentally changed. But this model is flexible so that you can assign view and edit rights as needed.

Tip   As we'll see in a moment, the description field on panels is displayed alongside the panel name. While you can't set this when adding a panel to the library, it can be set in the panel options. It's a good idea to add a short description to panels that are going into your library to help people understand what they are and how they're used at a glance.

Using a Library Panel in a Dashboard

Adding a library panel to a dashboard is simple. Start by clicking the add panel button, just as if you were creating a panel from scratch. But instead of adding a new panel, select "Add a panel from the library." This will bring up the panel library list, as shown in Figure 7-15, which allows you to filter and search through the library for the panel that you're looking for.

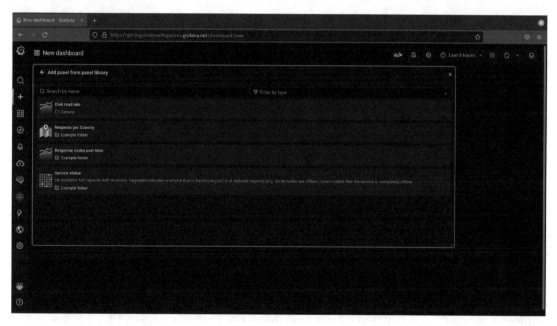

Figure 7-15. *Selecting a panel from the library*

Note in Figure 7-15 that you can see panel descriptions as well as names in this list, meaning that it's a good idea to set a description on a panel before adding it to the library.

If you have too many panels in the library to quickly find what you want, you have two options. You can use the search box on the top left to search for a panel based on its title. (This is a great reason to add prefixes to panel titles when adding them to the library, as they can be found quickly and will be grouped together in this view.) You can also filter based on the type of visualization, so if you remember that the panel that you wanted was a bar chart but can't remember the name, selecting the bar chart type will filter out all other types of visualizations.

Managing Library Panels

Library panels can be searched, viewed, and deleted in an interface similar to that of dashboards, as shown in Figure 7-16. To find the library panels management view, open the Dashboards menu from the navigation bar on the left and select the "Library panels" menu item.

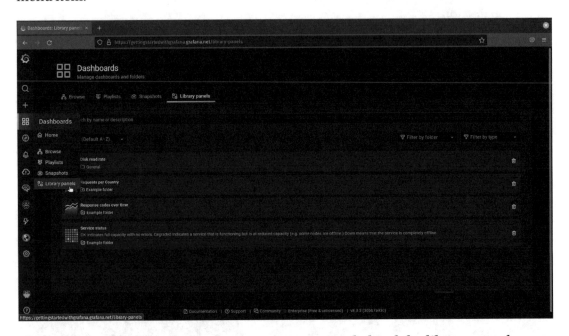

Figure 7-16. *The library panels management view behind the library panels management entry in the Dashboards menu*

In the library panels management view, you can search for specific panels using the same filters as shown when adding a panel to a dashboard. Here, though, you have two more options.

First, you can remove panels from the library by clicking the delete icon to the right of each panel's name. You can't remove a library panel that's being actively used as this would break existing dashboards, so if you try to remove a panel that's in use you'll be shown a list of dashboards that currently have this panel added. To delete the library panel, first remove it from these dashboards and then click the delete icon again.

The other option provided by this view is linking you to all dashboards that are using a given library panel by clicking that panel's name. Figure 7-17 shows a view of a library panel that is being used by two dashboards. Selecting one of those dashboards and clicking the "View panel in dashboard" button will take you to that dashboard.

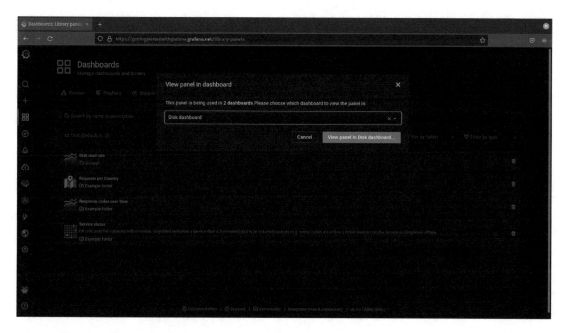

Figure 7-17. Viewing a panel in a dashboard

Updating Library Panels

By design, library panels are meant to be the same on every dashboard. That means that if you make a change to a library panel on one dashboard, it will be updated everywhere else that panel is used.

In order to update a library panel, start by navigating to a dashboard that contains the panel. If you don't know which dashboard contains the panel, you can either create a new temporary dashboard and add the library panel to it or navigate to it through the library panels management view.

Once you've finished making changes to the library panel and navigate away, you'll be prompted as to whether you want to save these changes to all library panels, as shown in Figure 7-18.

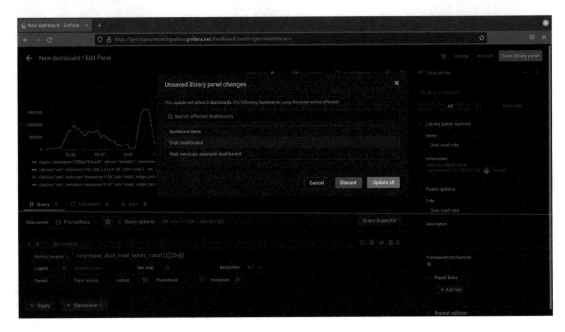

Figure 7-18. *Updating a library panel*

If you select "Update all," the library panel will be updated, and all dashboards using that panel will now receive the changes you made. If you don't want to update all the dashboards that use this library panel, select "Discard" instead.

Unlinking Library Panels

There are times when you want to make a change to a library panel in one place, but not update that panel everywhere it exists. For example, if you're working on an updated version of the panel but need to test some things out, you might want to work on a copy until you have everything completely ready to update. Or you might want to use an existing library panel as a starting point for a totally different panel.

In order to make changes to a single instance of a panel and not update the whole library, you can *unlink* that panel from the library panel. In effect, this will make a copy of the library panel on your dashboard, but it will no longer be attached to the original library panel. Any changes you make to it now will take effect only to your isolated copy.

To unlink a panel from the library, start with a library panel on a dashboard. Open the panel menu and the "More..." submenu and select "Unlink library panel" as shown in Figure 7-19.

Figure 7-19. *Unlinking a library panel*

Once you select this item, you'll be prompted to confirm unlinking the panel. Once you confirm this, your panel will no longer be connected to the library and can be edited without affecting other dashboards.

Caution   Unlinking cuts both ways! Your changes won't update the library, so they won't affect other dashboards. But if the library panel is later updated, your unlinked panel won't receive those updates. Be sure this is really what you intend before unlinking a panel.

Summary

In this chapter, we've explored the concept of a workflow in Grafana. You've seen how to create links from panels and from data points to other dashboards or external systems. You've learned how to use the explore view to quickly query data sources and to look at multiple queries side by side. Finally, you've seen how to create a library of reusable panels that can be managed and updated centrally.

In the next chapter, we'll look at extending your dashboards and panels to work with data from multiple sources simultaneously. You'll learn how to mix data sources together effectively and even how to derive new data by combining and comparing multiple sources.

CHAPTER 8

Working with Multiple Data Sources

So far, the dashboards and panels we've looked at have contained data from a single source. But there's rarely a single source of truth in any environment of any real size or complexity. Even when organizations decide to standardize on a single data storage solution, that migration can take months or years during which data is split between the new system and the old ones. Some systems may not be able to move at all due to regulatory issues, lack of knowledge of the underlying infrastructure, or simply because it's not worth the effort to move a functioning system to a new environment. And often when a migration is finally complete, it's taken long enough that the new system is now out of date, and it's time to start yet another migration!

In the real world, you'll always need to be able to correlate and visualize data from multiple sources. And being able to connect business data to IT and infrastructure data can let you derive new insights and take action: understanding how much money you lose per minute when the latency in your ecommerce site hits a critical threshold means that you can make a solid case for the budget to scale it appropriately!

Grafana provides a wealth of tools for visualizing data from multiple sources in the same place, both within a single dashboard and combining different sources of data in a single graph.

Side by Side

When exploring design principles in Chapter 6, we briefly looked at presenting data from multiple sources in a single dashboard as separate panels. Let's dig in deeper and see how to make this effective.

© Ronald McCollam 2022
R. McCollam, *Getting Started with Grafana*, https://doi.org/10.1007/978-1-4842-8309-7_8

Keeping Things Consistent

Whenever presenting multiple sets of data, your most important task is to not mislead people about what the data actually represents. This is pretty straightforward when showing a single data source, assuming that your data is correct: throw a query into a panel and Grafana will generally do the right thing automatically.

One of the things Grafana will do is to automatically set boundaries on many visualizations in order to keep them tightly focused on the data. Normally, this is a good thing, as it shows the data in the largest space and leaves out unused area. But when you're comparing similar data with different ranges, you need to be aware of the fact that your data might not always line up perfectly.

Consider Figure 8-1. In this example, we have two sets of temperature data, both in Fahrenheit. But the ranges that the data fall into are not the same. One location is much warmer than the other. If you use the Grafana defaults, you'll get a set of graphs like the top two, which implies at first glance that the two locations are roughly the same temperature – only after looking at the Y axis would you notice that the range of values is quite different. This is a misleading visualization and can lead to confusion or poor decisions. It's much better to ensure that the graphs have a consistent range, as shown in the bottom row of Figure 8-1. In this case, it's immediately obvious that the two locations don't overlap in temperature at all.

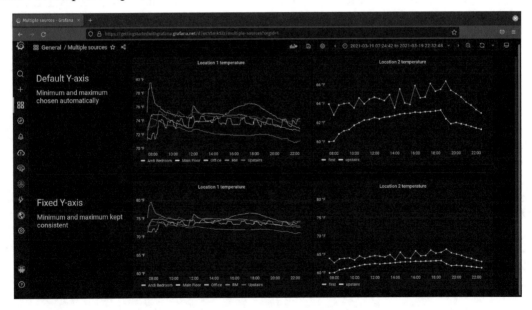

Figure 8-1. *Automatic Y axis ranges (top) can be misleading when comparing data. If ranges can differ, set the Y axis range manually (bottom)*

To set the minimum and maximum ranges of the Y axis, look in the panel settings under "Standard options," as shown in Figure 8-2. You'll need to make sure these are consistent across all the data that users of your dashboard will compare, so try to capture the full range of expected values when setting these.

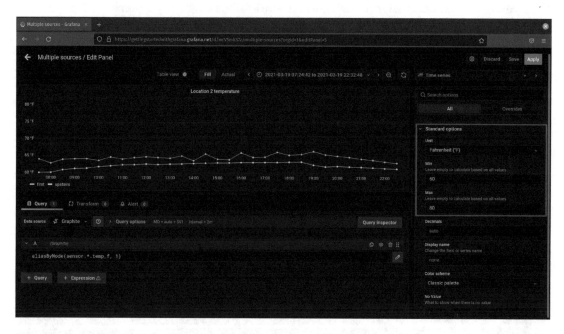

Figure 8-2. *Setting the Y axis range in the panel options*

(Also note that the unit is set here; as discussed in Chapter 6, it's always best practice to set this field. Even when comparing data with the default thresholds, we at least knew that the temperature was in the same units for both graphs!)

Caution   Be sure to consider extreme cases as well as typical use when setting your ranges. If you make your range too narrow, data that falls outside of it will not display on the graph. But if you make your range too broad, small changes will not be easily visible. Consider looking at a large time range of your data in the explore mode to get a feel for both the normal range and any outliers.

Staying in Sync

Having two panels side by side is great for small sets of data. If you can have two or three panels in a row, it's easy to see how the data lines up. But often you'll have more data than you can reasonably fit into a single row without making the panels too small to see.

In cases where you want to be able to line up data points across many panels or panels that are spread out, you can enable a shared cursor across all visualizations on your dashboard. Figure 8-3 shows multiple panels sharing both a cursor and tooltips across multiple panels. Notice that the mouse cursor is hovering over a data point on the panel in the upper right, showing concurrent total users logged in to a service. Other panels show a crosshair indicating data for the same point in time. In this case, we're also showing the actual data point itself in a tooltip, which can be useful information but can also add visual clutter if there are too many panels.

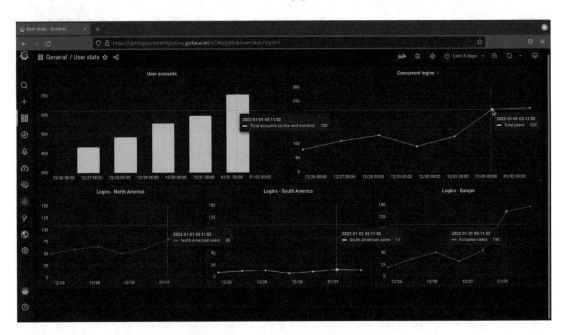

Figure 8-3. *Multiple panels showing data for the same point in time, highlighted with shared cursors and tooltips*

The cursor and tooltips will update in real time as the user moves their mouse cursor around the dashboard. This makes it easy to see which points line up when graphs are far apart or there are too many to compare easily. (Note also that while the three regional breakout panels at the bottom all share a Y axis range, the total grouping in the top right has a different range. Using a shared cursor makes it apparent how this data lines up. If instead all of these panels used a single range, the rapid growth in European users in the bottom-right panel might not be as apparent.)

Shared cursors and shared tooltips are enabled at the dashboard level rather than in individual panels. In order to turn this setting on, click the dashboard settings icon in the upper right and select the type of cursor sharing that you want as shown in Figure 8-4. *Default* will show crosshairs and a tooltip in the currently selected panel, *shared crosshair* will show the crosshair cursor in all panels but tooltips will appear only in the currently selected panel, and *shared tooltips* gives both the cursor and tooltips for all panels on the dashboard at the same time.

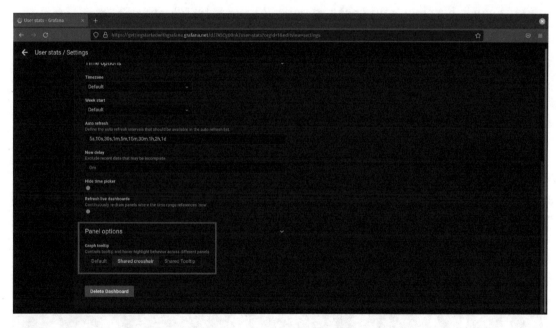

Figure 8-4. *The dashboard settings view with the cursor and tooltip options highlighted*

In-Panel

Grafana isn't limited to one data source per graph. You can incorporate multiple different queries from as many data sources as you like within a single panel with just a few extra clicks.

The "Mixed" Data Source

Normally, when creating a dashboard panel, you select the data source that you want to query and then add one or more queries to that data source. In order to use multiple different data sources, you need to select a special data source called *mixed* for your panel. This tells Grafana that you want to mix data sources in a single panel.

Tip   The mixed data source appears at the end of the list of possible data sources, so if you have a large number of data sources, you'll need to scroll to the end to see it. It's always present and doesn't have to be added to Grafana ahead of time for you to use it.

As you'll see in Figure 8-5, selecting the mixed data source adds a new field next to each query that you are running. When you click the "+ Query" button, you'll have the option to set the data source that you want that specific query to run against. Each data source can have a different syntax for running queries, so you will likely see differences in the query window for each data source that you use.

Figure 8-5.   *Editing a panel with multiple data sources. The panel data source is set to "mixed," and each query has its own data source attached*

Dealing with Multiple Axes

There's a point to note when adding additional queries and data sources to a panel. By default, Grafana will assume that all the data shares the same units and should be put on the same Y axis. If this is the case for your data, then there's nothing additional you need to do. For example, if you are looking at sales numbers in US dollars that are stored in different databases, there's nothing additional that you need to do.

But if you want your weather dashboard to compare sales to the number of visitors to your website, you need to have these called out separately. It certainly makes sense to compare those values as they are interrelated, but they use completely different units and scales – you can't sensibly talk about measuring sales in numbers of clicks or measuring users in dollars.

Fortunately, Grafana provides *overrides* which give you the ability to add additional Y axes and to change default settings on a per-series basis. We'll take a much deeper look at overrides in the next chapter, but let's take a look at controlling the Y axis now.

In Figure 8-6, you'll see a time series panel representing temperature (in degrees Fahrenheit, stored in Graphite) and relative humidity (expressed as a percentage, stored in Prometheus). The option is set to show the scale for temperature, but that doesn't really make a lot of sense when adding pressure into the visualization. So in this case, you would need to add an override to change how Grafana displays part of the data.

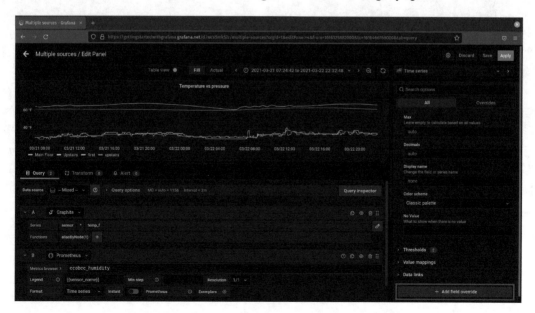

Figure 8-6. *Multiple data sources on a single graph, but without proper overrides applied. The "Add field override" button is used to change this*

There are many ways to specify what exactly you want to override on a graph. In this case, we want to change everything that is coming from the Prometheus data source, which Grafana has helpfully labeled as query "B." So selecting the option to override all fields from query "B" will easily give us what we need.

Once you've selected a source for overrides, you can select specifically what you want to change from the default behavior. Any option that is available to set on a panel can be overridden, which we'll explore in the next chapter.

In Figure 8-7, you'll see a set of overrides selected to change the behavior of the humidity data. The axis placement for this series is set to "right," meaning that while normally data will be graphed against a left Y axis, for this specific set of results it will be graphed against a second axis on the right. The unit option specifies what unit will replace the default of degrees Fahrenheit. In this case, that's a simple percentage. Finally, a change to the line style to represent humidity as dashes clearly separates them from the temperature lines.

Figure 8-7. *A set of overrides applied to one set of data. The resulting graph shows a clear relationship between humidity and temperature*

Calculating New Data

Most of what Grafana is used for is visualizing information directly – taking a series of data and presenting it visually in a way that can be more easily understood by the viewer. The bulk of the options that Grafana provides are for doing this sort of formatting or representation of data.

But sometimes you need to make changes to your data before visualizing it. For example, you might have temperature readings coming from several different data sources in different units. Some might be in Fahrenheit, some in Celsius, and some might even be in Kelvin. Meaningfully representing all of these in a single visualization is impossible without converting all of them to the same unit first.

Fortunately, Grafana has a few tools built in for performing mathematical operations on your data before it's visualized.

Tip   The best place to make changes to your data is always in the data source itself. If your data source has the option to perform a mathematical operation or to rearrange your data before it's sent to Grafana, you should always do it there – it's almost certain to be faster and more feature-rich than using Grafana for this, especially as much of the processing that Grafana does happens in the user's browser. If you have a SQL database, as an example, doing your calculations in your SQL query is definitely preferable to doing the same calculations in Grafana. The functionality that Grafana provides for this is useful, but should be considered a last resort!

Math with Expressions

Grafana *expressions* are a way of manipulating data that has been returned by a query. Expressions apply to an entire set of data, whether that's the result of a query directly or a set of numbers and operations that have been input. They are primarily intended for managing data sets to be provided to Grafana's alerting system, but can also be used to alter data before it is visualized. As a result, expressions are run server-side, and the data is processed on the Grafana server before being sent to the user's browser.

We'll be looking at a single expression type here in order to do simple calculations. There's a lot that expressions can do that aren't covered here. For more details, consult the Grafana documentation on expressions: *https://grafana.com/docs/grafana/ latest/panels/expressions/*.

Adding an expression to a graph is similar to adding another query. In fact, once added the expression will behave as a query. It adds data to the visualization and can have overrides applied like any other series. You can even apply an expression to the results of an expression to perform more complex tasks.

To add an expression, start by clicking the "+ Expression" button in the panel editor. In this case, we want to do a simple mathematical operation to change Fahrenheit to Celsius, so select the "math" operation for your expression.

The expression itself is a mathematical expression. You can reference other series by using a dollar sign ($) followed by the series name. Figure 8-8 shows referencing the first series, called "A" by default, and applying the formula to change it from Fahrenheit to Celsius. A new series is created, applying the formula to each data point in the original series to create a new series. By default, both the original values and the expression will be shown, but you can always hide the original data by clicking the enable/disable query button, which is the eye-shaped icon to the right of the query name.

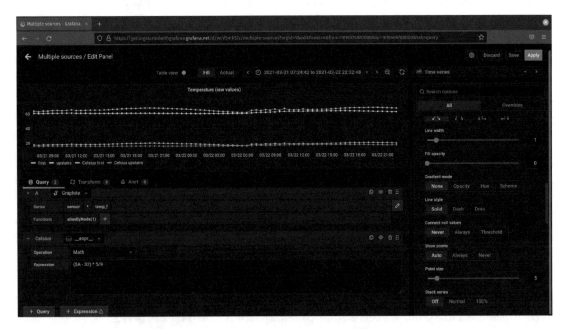

Figure 8-8. *An expression applied to a data set, converting Fahrenheit to Celsius*

> **Note** In this case, there is no unit set, so the temperature readings appear to be different. If you follow best practices and set the appropriate units for each set of data, Grafana will properly account for the differences between Fahrenheit and Celsius on the graph, and the data sets will look exactly the same. However, in this case, since the lines would match exactly, it would be difficult to see the results of the expression.

The mathematical expressions that you can create are not limited to a single series. You can apply complex formulas between multiple series by using their names in your formula, exactly as $A is used here. Just remember that for the expression to work, you need to have data points that overlap and contain the same time readings.

Math with the Add Field Transform

Transformations provide a more flexible and powerful way of manipulating data. Unlike expressions, transformations are intended to alter data for visualization purposes only. As a result, they provide a number of mechanisms for changing the look and feel of a data set when it is being graphed.

We'll look into transformations in more detail in the next chapter, but there's one transformation in particular that is especially useful when working with multiple different data sources: the *add field from calculation transform*.

As the name implies, the add field transform adds additional fields or data sets to your existing data by performing some sort of calculation on it. This can include things like finding the average, minimum, or maximum of a data set, looking at the delta between the start and end data or between every point in the series, and much more. But it can also be used to quickly relate multiple series for simple calculations.

To use a transformation, select the "Transform" tab in the panel editor, which is next to the default "Query" tab. To use the add field transform, select it from the list. This will present a set of options that can be used with this particular transform.

To do simple mathematical operations, select the "Binary operation" mode. This will enable you to perform math on two fields or numbers. Figure 8-9 shows an example of a binary calculation comparing the readings of an indoor temperature sensor to an outdoor sensor. The mathematical operator here is subtraction, so the minus sign (-) is selected between the two series being compared.

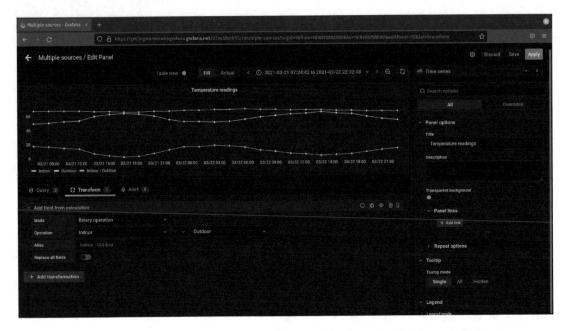

Figure 8-9. *Comparing two sensor readings with the add field transform*

Just like with the expression we used earlier, the add field transform adds a new series of data to the graph, this time indicating the delta between the indoor and outdoor temperatures. But unlike with expressions, we can't hide the original data by using the enable/disable query button. Because transformations run in the user's browser after Grafana has run a query, hiding that data in the panel will also hide it from the transformation. If you do hide one or both of the original queries, your new series will also disappear.

There are other ways to show or hide data on panels using transformations that we'll look at in the next chapter, so for now just keep in mind that transformations work on what you as a viewer can see, so you need to make sure your data is always visible to them for them to work.

Summary

In this chapter, we looked at two different techniques for working with data from multiple sources. When visualizing data in separate panels, you can align panel options so that data is consistent and meaningful across all visualizations. You saw how to sync up the cursor and tooltips across multiple different panels within a dashboard. We previewed some more advanced override functionality to see how to apply appropriate units and axes to data resulting from multiple queries.

You also learned how to apply mathematical expressions to data retrieved from multiple sources of data, both server-side on the Grafana server with expressions and client-side on the viewer's browser with a transformation.

In the next chapter, we'll look at these topics in more depth. You'll learn how to use variables in dashboards to allow them to be customized without being edited, and we'll look at more functionality provided by overrides and transformations.

## CHAPTER 9

Advanced Panels

So far, we've looked at working with data directly in panels and arranging those panels on dashboards. Now we can take things a step further and see how to customize visualizations in more detail.

In this chapter, we'll look at overriding default behavior for visualizations for specific portions of data while leaving the rest unchanged. And we'll look more deeply at transformations, learning how to reformat and alter the data itself.

There's a lot to cover here, but it all builds on material previously covered. You'll find that as you explore more of Grafana's functionality, the things you've learned will continue to serve you – we'll just be going deeper into the functionality provided and exploring beyond the defaults.

Panel Overrides

Panel overrides let you change options set in a panel for specific portions of the data that's being visualized without changing the global settings for that panel. In effect, they override the settings that have been chosen as the standard ones for that panel.

We've seen overrides already in Chapter 8 when working with multiple series of data in the same panel. In that example, we used an override to specify that a set of data should use a second Y axis.

Adding Overrides

Overrides are created by clicking the "Add field override" button at the bottom of the panel options section when editing a panel. Once you click this, you'll see that there are multiple ways to choose what values are being overridden. Figure 9-1 shows the options for selecting fields, which we'll review as follows.

© Ronald McCollam 2022
R. McCollam, *Getting Started with Grafana*, https://doi.org/10.1007/978-1-4842-8309-7_9

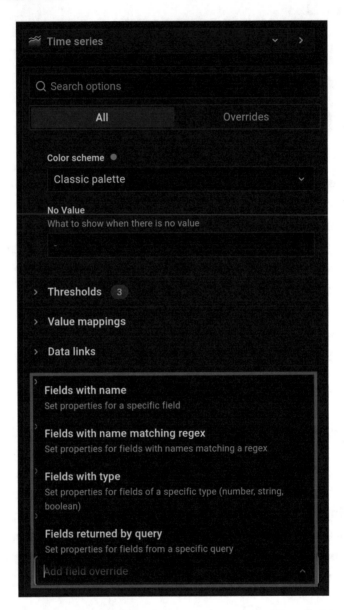

Figure 9-1. *There are several options for selecting the field or fields to apply overrides to*

Fields with Name

This option will present you with a list of all of the series of data presented in the panel based on their names. It's most useful if you have a single series (or small number of series) that you want to apply overrides to individually. It's also very specific, so if the data returned from your query changes in a way that makes the name of a series change, your overrides will no longer apply. It can also be tedious to look through a large number of series names, so if you have more than a few items in your query results, consider using one of the other methods.

The selection will match the name of the field shown in the graph legend, so if you have set custom names or made changes in the legend, these will be what you see in the selection box, as illustrated in Figure 9-2.

Figure 9-2. *Any changes made to the legend for a series will be reflected in the list of series names for overrides*

Fields with Name Matching Regex

Similar to the "fields with name" option, this will let you choose the series to override based on the name. However, rather than picking a single series name, this option lets you use a regular expression to choose which series to override.

Using a regular expression gives you a few benefits over the "fields with name" option. It can select multiple series at once, letting you apply overrides to many data series at the same time. And if your regular expression is crafted well, it can handle cases where the names of series might change over time.

In order to use this field selector, type or paste a regular expression into the regex field. You can optionally surround your regular expression with slash (/) characters, but this is not required. So both /95.*/ and 95.* would select the "95th percentile" series in Figure 9-2.

Fields with Type

The "fields with type" selection is a bit different. Rather than choosing a series based on its name, this allows you to choose all series with a given type of data. After selecting this option, you'll see a dropdown box with all of the types of data available in your graph. The types available will depend on the data that you have. If you have numeric time series data, for example, you'll be able to select numeric or time values. But if your data is tabular without timestamps, you might see string and numeric options but not time.

This override type is most useful when you want to manage an entire category of data. For example, Figure 9-3 shows a table with overrides applied to only string data and not numeric data in order to make row headers stand out more and to replace empty values with a human-readable string of "Undefined." (Note that this applies multiple *override properties*, which we'll look at next.)

Figure 9-3. *Setting override properties on string values does not affect numeric values*

Fields Returned by Query

This field selector applies to all data that is returned by a specific query in this panel. It's most useful when you are comparing two or more *sets* of data to each other rather than adjusting the visualization of a single series or data type.

This can be used to compare data with different units of measurement over the same time period. For example, if you are looking at temperature and pressure but measuring each of these in multiple locations, it can be handy to apply a set of override properties to all of the pressure measurements at once. Or, as shown in Figure 9-4, when comparing CPU and disk utilization for the same system that has multiple disks and multiple processors, it can be handy to apply overrides to an entire set of data returned by one query.

When using this selector, you'll have the option to choose which query's data to override. This corresponds to the query name in the section on the left. Figure 9-4 highlights the query selected for clarity.

Figure 9-4. *Applying an override to a query will change the visualization for all data returned by that query*

Tip   Grafana will automatically name your queries as you add them. By default, these will simply be letters of the alphabet, but you can change the query names to be more descriptive by clicking the default name and typing a new one. This can be especially helpful in panels with lots of queries or when applying complex overrides – it's far easier to see that applying a "degrees Celsius" unit to a query is the right thing to do when it's named "outdoor temperature" instead of "C"!

Using Override Properties

Once you've selected the fields or queries that you want to make changes to, you can start applying *override properties* to them. Override properties are changes to the defaults that you've selected for visualizations in your panel. Effectively, these tell Grafana how to change the display of the specific data that you've selected from the rest of the data in the panel.

To add an override property, click the "Add override property" button inside the override that you've created. When you do this, you'll be presented with a list of possible properties that can be overridden, as shown in Figure 9-5. This list can be quite long, as some visualizations have a lot of properties that can be set! So rather than scrolling through the whole list, you can type part of the name of the property to filter the list down. If you can't remember the name of the property that you want, just scroll up a bit and look through the panel options; the name will be the same in both places.

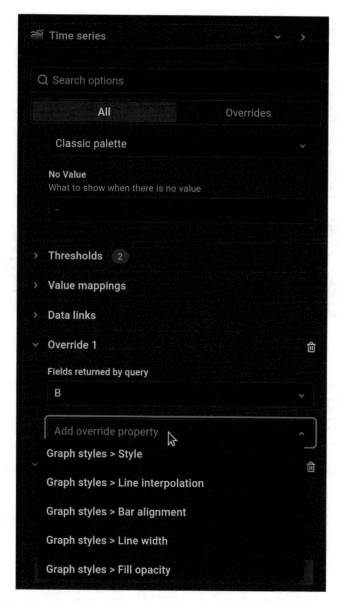

Figure 9-5. *Adding an override property lets you change any property that can be set for the visualization*

You can set more than one override property within a single override. Figure 9-6 shows an example of this, with a single override for query "B" changes properties for the axis placement, units, line type, and line thickness. This lets this chart show the relationship between temperature and humidity for a time period in a sensible way.

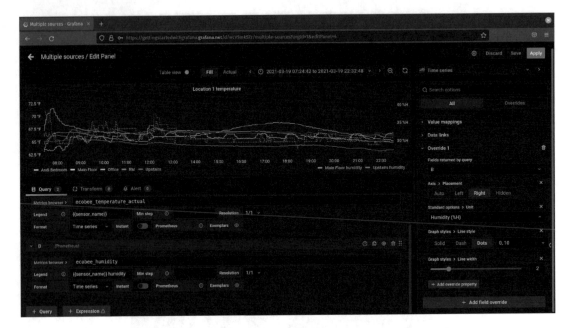

Figure 9-6. *Multiple properties can be set within the same override*

Note   The options that can be overridden depend on the options presented by the original panel. For example, the pie chart panel lets you select whether to use a full pie or a donut chart as your visualization, which doesn't make any sense when using a bar graph. So the pie/donut option will appear as an override for the pie chart but not the bar chart.

Some panels don't have any options that can be overridden. In this case, you won't see the option to add an override at all.

Removing Overrides and Override Properties

To remove an override property from an override, click the small x to the right of the override property entry. This will remove the property from the override, but leave the override itself in place.

To remove an override completely, click the remove override button with an icon of a trash can to the right of the override name. This will remove the override entirely, including all the properties that you've set in that override.

Transformations

Transformations are a powerful feature of Grafana that allow you to combine, change, and otherwise manipulate data after it has been retrieved from a data source but before it is visualized. Think of them as an extra processing step that sits between a query and actually graphing your data. For example, transformations can allow you to perform mathematical operations on raw data where the data source doesn't support it. They can even let you perform SQL-like joins on data from multiple different sources!

There are a few things to keep in mind when using transformations in Grafana. First, because they take place after the data has been queried from the data source, all of the work that transformations do happens in the user's browser. This means that very large numbers of transformations or transformations that run on very large data sets might slow down the browser of everyone who views your dashboard. It's always better to **do as much as possible in your data source** – keep transformations for the things you can't do there or for light formatting whenever you can! A large database server will run joins much faster than your web browser.

Also bear in mind that transformations take place before the visualization plugin gets access to your data. Whatever is changed in your transformation is what the panel will get to work with, and it will not have access to the original data at all. So, for example, if you hide a field in a transformation, that field disappears from your data entirely when it is graphed. You can't use a transformation to hide a field that you want to sort on in your panel. And it's entirely possible to use transformations to change your data enough that it can't be graphed. If you hide the time field from your data, the time series panel won't be able to display the data at all, so you'll need to use a table instead.

Finally, you need to know that transformations run in order. You can string multiple transformations together one after another, and each transformation receives the results of the transformation before it. This means that changing the order of transformations can change the data entirely. When using multiple transformations, be sure to think about the order carefully.

With those points in mind, we're ready to start exploring the power of Grafana's transformation system!

Managing Transformations

Transformations are managed from the panel edit view. To work with transformations, start by editing a panel and retrieving some data from a data source. You can then click the "Transform" tab under the panel preview to access the transformations, as shown in Figure 9-7.

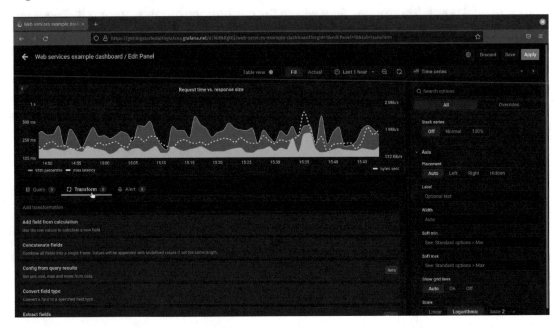

Figure 9-7. *Grafana transformations are accessed by selecting the "Transform" tab when editing a panel*

To add a transformation, select it from the list. You can use the search bar to quickly find a transformation if you know the name already.

Each transformation has its own set of options. We'll review some of the more commonly used ones later. In most cases, you'll need to fill out the required options for the transformation to take effect. Once you have completed this, the transformation will immediately occur, and your panel preview will update. You can add more transformations by clicking the "Add transformation" button.

Once you've added a transformation, some additional options will appear. Each transformation has a set of possible operations, as highlighted in Figure 9-8. From left to right, these buttons provide a link to help for the transformation, enable debug mode, enable or disable the transformation, remove the transformation, and allow you to reorder transformations. We'll briefly examine each of these.

Figure 9-8. *Transformation menu items apply to each transformation individually*

Transformation Help

Clicking the transformation help icon will show a brief description of the transformation and provide a link to the Grafana documentation for this transformation.

Debug

The debug icon will open a special view that shows the data before and after the transformation runs, as shown in Figure 9-9. This view doesn't allow you to edit or change the data in any way; it's specifically for examining the effects of the transformation. It's something you shouldn't need to use under normal circumstances, but if you ever become confused as to what's happening under the hood in a given transformation, this view can help you understand what's going on at the lowest levels.

Figure 9-9. *The transformation debug view can help you understand why a transformation produces its result*

Enable/Disable Transformation

Transformations are on and run as soon as they're created by default. This button will turn a transformation off so that it doesn't take effect but without deleting the transformation altogether. Clicking it again will cause the transformation to turn back on. Enabling and disabling transformations can be useful when debugging longer flows of transformations.

Remove Transformation

The trash can icon will remove the transformation from the panel. This won't affect other transformations, so if there are transformations before or after this one, they'll still take effect.

Move Transformation

Dragging this handle will let you move the transformation up and down in the list. Transformations run top to bottom, so transformations above will run before the ones below them.

Commonly Used Transformations

The transformation feature was added to Grafana in Grafana 7 and, as of the time this book is being written, is still developing quickly. There is already a long list of available transformations, and more are being added on a regular basis, so it won't be possible to cover every possible transformation here. In this section, we'll take a look at some of the most commonly used transformations, but you can always see the full list in the Grafana documentation at *https://grafana.com/docs/grafana/latest/panels/ transformations/*.

Tip   If you want to experiment with a transformation but don't have the exact data you'd need, try using the Testdata DB data source from Chapter 4. You can use this to generate various types of random data or even use the "CSV content" setting to allow you to mock up your own data in CSV format.

Add Field from Calculation

This transformation is used to create new series of data based on one or more existing series from your data source. It applies a modification to the original data to create a new field that can be graphed.

The add field transformation has two modes, *reduce row* and *binary operation.*

Reduce row will take one or more of your fields and apply a reducing function such as taking the mean, the count of values, the sum total of all values, etc. By default, this will apply to all of the fields in your visualization, but you can select specific fields to use in the calculation by clicking their names. Figure 9-10 shows an example that uses this transformation to add the mean of two separate series as a new series in a visualization. (It also has a series override applied to make this new series a dashed line rather than solid so that it's more easily visible on the graph.)

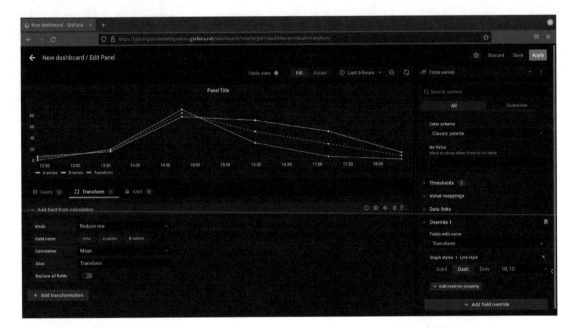

Figure 9-10. *A reducing function in an add field transformation can be used to graph the mean of data from two different sources*

When used in the binary operation mode, this transformation requires exactly two inputs to produce its new series. These inputs can be fields from your data or can be simple numbers. You must also select a binary operation to perform on these inputs, which is one of addition, subtraction, multiplication, or division. Figure 9-11 shows a binary operation calculating the difference between two series.

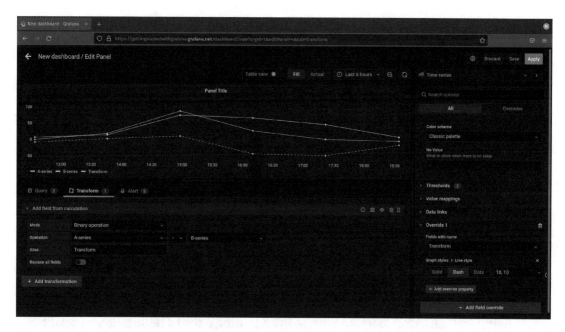

Figure 9-11. *The binary operation mode of the add field transformation allows simple mathematics on multiple sources of data*

In either mode, selecting the "replace all fields" option will hide the original data and display only the newly calculated values.

Extract Fields

The extract fields transformation is one of the most powerful tools available in Grafana. It allows you to do further parsing on data that is returned from a source in order to split compound fields (i.e., fields containing multiple different values at once) into separate fields for visualization. It can handle both JSON data and key/value data (data that looks like key1=value1,key2=value2).

Let's take an example of a manufacturing line that provides a timestamp and JSON data about each step in its process. Listing 9-1 shows what a sample JSON block from this system might look like.

Listing 9-1. The output from our example manufacturing plant data source

```
{
  "plant": "eu",
  "detail": {
    "assembly_step": 12,
    "line": 3,
    "component": "pick_and_place"
  },
  "state":"normal"
}
```

There's a lot of great detail here that we could use for things like showing status with the status history panel, flow with the state timeline panel, or even just the data in a nicely formatted table. But since this data source returns all of that information as a single field, Grafana can't do much with it. Even putting it into a table gets us a result like Figure 9-12.

time	metric
	Manufacturing status
2022-01-23 06:48:04	{"plant": "eu","detail": {"assembly_step": 12,"line": 3,"com...
2022-01-23 06:49:07	{"plant": "eu","detail": {"assembly_step": 13,"line": 3,"com...
2022-01-23 06:51:18	{"plant": "eu","detail": {"assembly_step": 14,"line": 3,"com...
2022-01-23 06:51:19	{"plant": "eu","detail": {"assembly_step": 12,"line": 4,"com...
2022-01-23 06:52:30	{"plant": "eu","detail": {"assembly_step": 13,"line": 4,"com...
2022-01-23 06:54:51	{"plant": "eu","detail": {"assembly_step": 14,"line": 4,"com...

Figure 9-12. *Even the table view isn't much use when all the information is in a single field*

By using the extract fields transformation, we can tell Grafana to parse the JSON data and pull the various parts into their own new fields.

The extract fields transformation will split out the JSON object into one Grafana field per JSON field. But note that this is a compound object – detail is itself another object with multiple fields. So to fully parse this data, we'll need to use the extract fields transformation multiple times, as shown in Figure 9-13. The first time, we'll use the "metric" field that's returned from the data source itself. But because this creates a new field with the parsed contents after the transformation, we can then tell the second extract fields transformation to look at the new "detail" field.

Figure 9-13. *The first use of the extract fields transformation adds the detail field, which the second use of the transformation fully parses*

Once this JSON information has been fully parsed, all the compound data is available to the visualization.

Group By

Group by lets you rearrange your data so that it's first grouped together by one or more columns and then lets you decide how to represent the other columns in your data. The group by transformation discards any columns that you haven't selected, so you need to explicitly select what to do with every column; any columns not selected in the transformation will not be part of the resulting data set. But by doing this, it gives you the ability to quickly reduce large, complex sets of data into something simple to work with.

Let's look at an example. Consider a set of sensor information from a smart home. This could include things like temperature sensors, a measure of the brightness of a room or area, and an indication of whether the lights are turned on or not. If this data is collected on a regular basis, you'll end up with a number of time series that can be graphed. But sometimes it's more useful to know some quick statistics rather than the full set of data, like the current state of all the lights and the range of temperatures throughout the day.

The group by transformation lets you choose what to do with each column, as the example in Figure 9-14 shows. First, select the column or columns you want to group your data into, then decide what to do with each other column. In this example, we're looking at the last (or current) state of the house lights, the maximum brightness, and the minimum, maximum, and average temperatures.

Figure 9-14. *The group by transformation lets you choose columns to use as buckets to group data into and perform calculations on other columns*

Once the transformation is applied, you'll see that your data has been bucketed into groups based on the column or columns that you have selected. Assuming that you have at least some duplicate values, you'll have a much smaller data set to work with, as Figure 9-15 shows. Both panels here are using the same underlying data, but the transformation has reduced the number of data points dramatically. Note that the columns are not the same on each, as the time column is ignored and several new columns are added in the transformation.

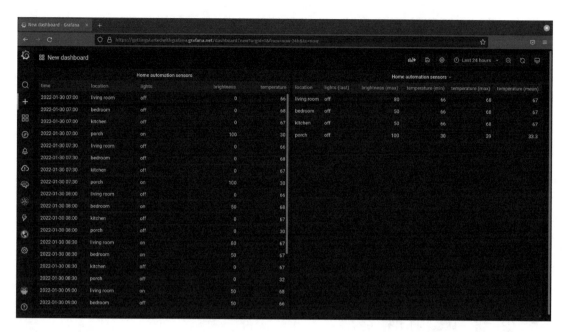

Figure 9-15. *A raw data set on the left and the resulting data set after applying a group by transformation on the right*

Tip   Grouping by a single column will sort all of your data into one bucket for each value that is in that column. Grouping by multiple columns will sort all of your data into one bucket for each unique set of values across all columns, meaning that you can end up with potentially many more buckets than you mean to. And if one of your columns has a unique value in every row, you'll end up with as many buckets as you have rows in your data set.

This set of unique values is referred to as the *cardinality* of the data set. Usually, when working with the group by transformation, you want to keep the cardinality as low as possible to reduce your data to workable sizes, so avoid using columns with values like unique IDs or timestamps as a grouping column.

Merge

The merge transformation lets you combine multiple sets of data into one set by automatically looking for overlapping values in identical columns in each data set. It won't have any effect on a single query that returns one set of results, but is useful to combine different queries that each contain a part of the final data you want to see.

For example, let's assume you have two data sources. One of these contains reliability data for data centers aggregated by fiscal quarter and tells you statistics like the uptime or reliability of each location and what its rate of successful responses to requests was, as shown in Listing 9-2. The other data source, shown in Listing 9-3, is also aggregated by quarter, but this one tells us the sales volume.

Listing 9-2. Quarterly reliability data for each data center

```
Quarter, Region, Reliability, Success rate
2021 Q4, US East, 99.88, 99.82
2021 Q4, US West, 96.07, 94.01
2021 Q4, Europe, 99.80, 99.88
2021 Q4, Asia, 98.28, 98.80
2022 Q1, US East, 99.61, 99.89
2022 Q1, US West, 99.03, 99.40
2022 Q1, Europe, 100, 99.85
2022 Q1, Asia, 99.02, 99.64
```

Listing 9-3. Quarterly sales data by region

```
Quarter, Region, Sales (thousands)
2021 Q4, US East, 140
2021 Q4, US West, 182
2021 Q4, Europe, 111
2021 Q4, Asia, 98
2022 Q1, US East, 144
2022 Q1, US West, 177
2022 Q1, Europe, 122
2022 Q1, Asia, 104
```

Unfortunately, both of these homegrown systems are in different places and don't support any SQL-like join functionality. Merge to the rescue! As long as the columns and values are the same in each data source, merge will automatically group all of the results from both data sources into a single table, as shown in Figure 9-16.

Figure 9-16. *The results of the merge transformation on Listings 9-2 and 9-3, giving a single table with all of the data*

Note   There are no options for the merge transformation – it detects all matching fields automatically. If you want to control which field is used to join multiple tables, check out the *outer join* transformation.

Organize Fields

The organize fields transformation is simple but one of the most useful. It lets you change the order of columns in your data, rename columns, or even hide them entirely. It's particularly useful when you have results that are meaningful but have cryptic names when they come from your data source. (It's much nicer to work with a field like `manufacturing errors per thousand units` rather than `mfg_err_k`!)

Figure 9-17 shows the organize fields transformation in use. The original field names are shown on the left, with the new names filled in the text boxes on the right (and reflected in the table preview above). You can change the order of fields by dragging the handle on the left of each field name up or down in the list.

Figure 9-17. *Using the organize fields transformation to make column names more readable*

Renaming fields can also do more than make them more readable to people, however. Remember that you can chain transformations together, so when using transformations like merge or outer join that rely on fields having the same name, you can use the organize fields transformation to make that happen.

Outer Join

Outer join functions in Grafana the same way that it does in a SQL database. It allows you to combine two data sets by matching identical values for a column in the data in both sets. This is similar to the merge transformation, but where the merge transformation will look for identical values across all columns in multiple sets of data, outer join works only on one field at a time. This means that with outer join, any other columns that are in both data sets will appear twice: once from each copy in the original data set.

As an example, consider Listings 9-4 and 9-5. These represent two sets of data, one with employee information and another with current assignments. We can use the outer join transformation to combine these into a single set of data in Grafana.

Listing 9-4. Employee information

```
emp_id, emp_name, emp_dept
12338, James Holden, Management
14586, Naomi Nagata, Engineering
15662, Roberta Draper, Security
16810, Amos Burton, Engineering
10674, Fred Johnson, Management
13440, Josephus Miller, Investigation
16906, Camina Drummer, Management
```

Listing 9-5. Employee assignment data

```
emp_id, assignment
12338, Rocinante
14586, Rocinante
15662, Scirocco
16810, Rocinante
10674, Tycho Station
13440, Star Helix
16906, Dewalt
```

In this case, the emp_id field is the same in each data set, meaning that we can use the outer join transformation to combine both sets of data in Grafana. Figure 9-18 shows the configuration of the outer join to use this field on both sets of data.

Figure 9-18. *Joining two sets of employee data using the "emp_id" field*

Tip   Unlike in SQL, the Grafana version of an outer join requires that the columns you are using have the same name. Remember that you can use the *organize fields* transformation before the outer join to rename one or more columns.

Reduce

The reduce transformation takes large or complex data sets and collapses them down to a small summary. There are two modes for this transformation: *series to rows* and *reduce fields*.

The series to rows form of the reduce transformation takes multiple series of data (the results of multiple queries or of one query that provides more than one independent set of data points) and performs calculations to reduce them to a single set of points, one for each field in the original data set. In this case, each field from the original data becomes a row in the new data set, and each calculation becomes a column.

Let's take an example of a set of temperature sensors in a restaurant, shown on a dashboard in Figure 9-19. Each sensor provides a reading at a regular interval, so there is a time series for each sensor which can be graphed as shown in the panel on the left. Applying the series to rows form of the reduce transformation lets you take a summary calculation for each field, creating a table with new values. The panel on the right shows the result of the transformation calculating the minimum, maximum, and average (mean) values for each sensor.

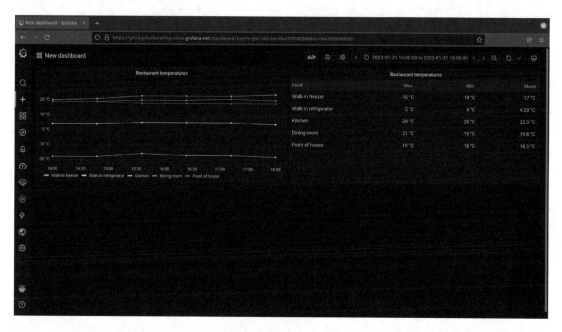

Figure 9-19. *The series to rows form of the reduce transformation allows easy calculations across multiple series of data*

The reduce fields option does not make any changes to the rows of the original series. There will be one row in the resulting data set for each row in the original, so in our restaurant example, we start with five rows and end with five rows. What this transformation will do is simply apply a calculation across each row of the original data. Figure 9-20 shows the result of applying this transformation to our restaurant temperature data using the last value, which will show the most recent temperature reading for each sensor.

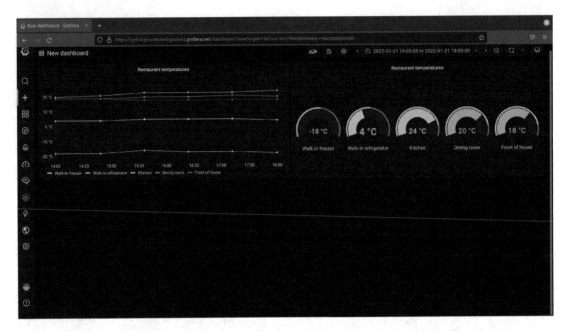

Figure 9-20. *The reduce fields form of the reduce transformation performs calculations for each row of data but leaves the number of rows unchanged*

Note   In this case, we could also have simply used the gauge panel type directly, as it already has an option to show calculated data such as the mean, max, min, or most recent value. The reduce fields option is mostly useful as part of a series of transformations where you need to collapse data to a single value before performing additional transformations. In general, if you just need to show one value and nothing else, it's easiest to have the panel do this for you rather than use a transformation.

Summary

In this chapter, you learned how to apply panel overrides, letting you specify a different presentation of data for specific fields or series of data. We then looked at the Grafana transformation system that enables you to manipulate data after it is queried but before it is displayed. We also reviewed some of the more commonly used transformations and saw how they can be applied to various forms of data in order to normalize data for comparison, reformat it for easier display and understanding, or reduce large data sets to a manageable level.

In the next chapter, you'll learn how to use dashboard variables to make your dashboards more dynamic and help you set up views for similar sets of data without having to build a large number of specialized dashboards for each data set.

CHAPTER 10

Dashboard Variables

So far, we've focused on building a view for every set of data that you want to see. Your dashboards have been useful and interesting, but static – each panel's data source and query has been set when you create it and doesn't change.

In this chapter, we'll move beyond this static representation of data and start to use variables to make your dashboards more dynamic.

Variables are a powerful feature of Grafana that allow a great deal of flexibility and dynamic control of your dashboards. Variables can be used to filter data quickly, carry information along from one dashboard to another when navigating in Grafana, and even as a way to easily repeat the same panel multiple times for a set of data without having to copy and paste that panel over and over.

We looked in Chapter 7 at how to add values to variables in data links and panel links to let us put data from a panel into a URL to be used by other dashboards or external tools. Now we'll explore how to use those variables in Grafana directly.

Managing Variables

Variables are managed from the dashboard settings view. After opening the settings, you'll see the *variables* tab on the left, as shown in Figure 10-1. In this view, you can add, remove, edit, and reorder variables.

© Ronald McCollam 2022
R. McCollam, *Getting Started with Grafana*, https://doi.org/10.1007/978-1-4842-8309-7_10

Figure 10-1. *The dashboard settings view showing the variable control panel. The variables tab is highlighted*

Like other similar settings panels in Grafana, the variables view shows all the variables that are currently set for a dashboard. Clicking the name of any of these will allow you to edit the variable definition. The trash can icon on the far right will delete an existing variable, and the duplicate icon immediately to the left of the delete icon will make a copy of the variable. The up and down arrows allow you to reorder variables on a dashboard – the variable at the top of the list will appear at the top left of the dashboard, and variables below it in the list show to the right in order when viewed on the dashboard.

The variable control panel will highlight potential issues for you where possible. For example, in Figure 10-1, the variables defined for the dashboard are not used in any queries, meaning they will show as options for the user to select but will not have any effect. This is indicated by the orange warning triangle next to the variable controls.

Variables that are referenced in queries but do not exist in the dashboard will be highlighted in the section at the bottom. These are important to note, as an empty or missing variable could cause a query to function incorrectly or even break entirely.

Defining Variables

Variables are powerful, but with that power comes some significant complexity. There are a lot of options and settings that can be defined for each variable, and they can even interact with each other dynamically. Rather than try to recap all of the Grafana documentation on variables (which is well worth reading!), we'll walk through some of the most useful functionality here and take a look at a few examples of how to use them effectively.

Let's start by creating a very simple variable to see how they work. Start by creating a new dashboard. Rather than immediately adding a panel, though, open the dashboard settings and select the variables tab, then create a new variable. You'll see a screen like Figure 10-2 which is where you'll set the options for your variable. There are a lot of options here, so for now let's focus just on the top section. These are the same no matter what kind of variable you're creating and need to be set (or at least intentionally left blank) for every variable.

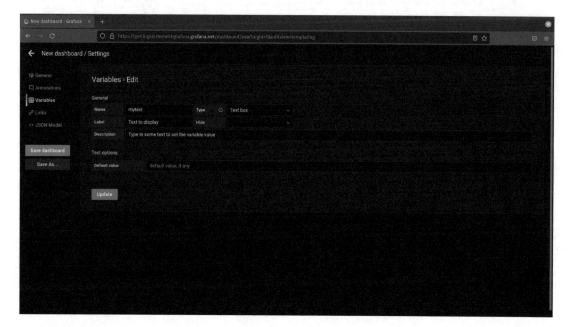

Figure 10-2. *The top section of the variable options is the same for all variables*

The fields here control what your variable is called and how users will interact with it:

- **Name:** What you'll use to reference your variable in queries. Think of this like a variable name in a coding language, so capitalization matters and only letters, numbers, and underscores are allowed (no spaces). It's also what users will see if you don't set the label field.

- **Type:** This tells Grafana where the values for the variable will come from. We'll look at this in more detail later in this section.

- **Label:** This is what will be shown to users on the dashboard. It's plain text, so you can use spaces or other special characters here.

- **Hide:** Controls whether the variable is shown to users or not. If left empty (the default), the variable is shown normally on the dashboard. Setting this to *label* will hide the variable's name or label, but still show a dropdown or text box to the user. Setting it to *value* will hide the variable entirely, which is useful when you want to use variables in URLs or for calculations, but not show them to the user directly.

- **Description:** The text entered here will appear as a tooltip when the user hovers their mouse cursor over the label. You can use this to provide additional details or instructions for using the variable.

To start, fill in the following:

- **Name:** `mytext`
- **Type:** `Text box`
- **Label:** `Text to display`
- **Hide:** `(blank)`

When you're finished, click the *update* button to save the variable and then the arrow in the upper left of the settings panel to return to the dashboard. You should now see a text box and your label at the top of your dashboard, as shown in Figure 10-3. You've now got a variable defined!

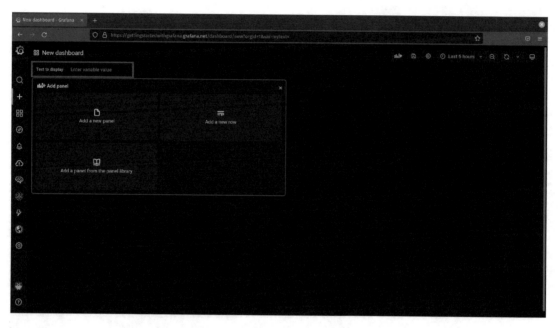

Figure 10-3. *A text box variable defined on a dashboard. The input field is created by Grafana and highlighted at the top of this image*

Of course, defining the variable is only half of the task. In order to be useful, you need to use this variable in your dashboard. Let's do this by using the text in a text panel. Add a text panel to your dashboard, leaving the default settings for Markdown text. But replace the content section of the panel options with

```
The value of your variable is:
# ${mytext}
```

Now when you enter text in the box at the top of the dashboard, it will appear in your panel. It will even update in real time as you change the text that you've written. Go on, try it out! You'll see the text you enter in your panel similar to Figure 10-4.

Figure 10-4. *Using a variable in a text panel is a quick way to see how variables can be used*

You can use this variable in any type of panel, not just text panels. For example, if you had a Prometheus environment with metrics from multiple different systems, you might use a time series or stat panel with a query like {host="${mytext}"}. Users could then type or paste a hostname into the text box to complete the query.

More Advanced Variables

Simple text boxes can be useful for free text searches, but they're annoying to use for many other types of data. In our example Prometheus query earlier, a user would need to type the hostname in exactly for the query to work – a typo or an extra character copied and pasted in would cause the query to fail. It would be far more convenient to allow the user to pick from a list of possible values here.

Let's look at some other ways to use variables in Grafana. For example purposes, we'll use the scenario of building a dashboard for a school to view student and class information. We'll use a single data source, in this case a SQL database with several tables, but remember that like everything else in Grafana, variables can apply to multiple different data sources.

In our example, we'll use a simple set of tables to represent students and course grades, as shown in Listings 10-1 and 10-2. The tables are related to each other by the `student_id` field, meaning that we'll need to use this ID to match up students to their grades.

Listing 10-1. Students in our example school

```
student_id, level, firstname, lastname
1, 3, Jane, Bloggs
2, 3, Joe, Bloggs
3, 3, Sara, Smith
4, 3, Cho, Kim
5, 3, Faaz, Ahmad
6, 3, Paarijata, Singh
[...]
```

Listing 10-2. Courses and grades in our example school

```
subject, student_id, score
Mathematics, 1, 89
Mathematics, 2, 87
English, 1, 84
Physical Education, 1, 88
English, 2, 91
Music, 1, 95
Music, 2, 78
English, 2, 91
[...]
```

It would be simple enough to represent all of this information in a Grafana dashboard. We could use table panels to show student enrollment, bar charts to show score distributions across courses and levels, and some simple SQL `JOIN` statements to show the results for the whole school.

But this will quickly become unwieldy for users. There might be hundreds of students at the school, each taking six or eight courses at once, meaning thousands of combinations of students and courses. An individual teacher would only want to see a few of these, and even administrators would want to be able to browse by a single course or level at a time. Fortunately, variables let us filter the data easily.

Custom Variables

After the text box variable type, the *custom* variable is probably the simplest. This variable type lets you define a set of possible values for a user to select. Let's take a look at a custom variable for selecting students from one of our tables. In Figure 10-5, you'll see a custom variable created called student. The values for custom variables are defined in the variable itself, so we can add values separated by commas (such as 1, 2, 3, 4, 5).

Figure 10-5. *Custom variables have values defined in the variable settings panel*

We can then add a table panel with a query to retrieve student IDs using this variable:

```
SELECT * FROM students WHERE student_id LIKE ${student}
```

Note   We're using the SQL LIKE keyword rather than the normal = here intentionally. Later on in this example, we'll extend these variables to be more useful and support wildcards, so we're just planning ahead a bit. If you're not familiar with SQL syntax, feel free to disregard this – it's not really directly related to managing variables in Grafana.

Now when the dashboard user selects a student ID from the dropdown at the top of the dashboard, they'll see only that specific student, as shown in Figure 10-6.

Figure 10-6.   *The value of the custom variable is passed into the query, replacing "${student}"*

This is a good start, but it means that we need to know the student ID that we're looking for. It'd be nice to use the student's name instead, but we need that ID number in the query.

Key/Value Pairs in Custom Variables

There's a great feature of custom variables that lets you show the user one thing, like a nice friendly name, and pass your query something else, like a hard-to-remember ID number. To use this feature, you can replace your simple values in your custom variable with key/value pairs. These are the friendly view and the value, separated by a colon (:). Note that there needs to be a space on either side of the colon! Without it, Grafana will assume that the colon is just part of the variable value. In Figure 10-7, we've replaced the simple values from before with key/value pairs showing student names and IDs.

Figure 10-7. *Key/value pairs in a custom variable will show the key (as previewed at the bottom of the page) but provide the value to queries*

Now when a dashboard user selects a student, they can do so by name, as in Figure 10-8. The SQL query will still get the student ID, so everything functions as before, just a bit friendlier.

Figure 10-8. *When using key/value pair variables, the selection box at the top shows the key, but the value is passed into the query*

This is much better, but still awkward to administer. Every time a student joins or leaves, we need to update the list of variables manually. It would be far better to be able to pull the student information directly from the database.

Query Variables

Rather than using a custom variable, let's change our student variable to a query variable. When you do this, you'll see a new set of options appear as shown in Figure 10-9.

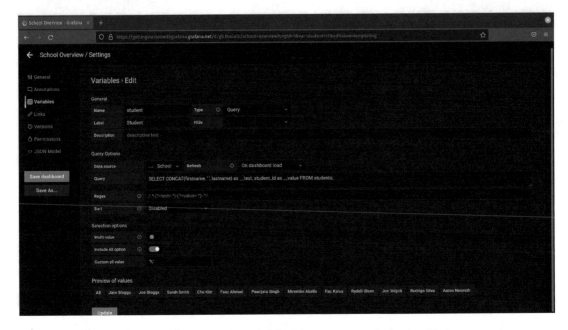

Figure 10-9. *Selecting the query variable type exposes more options*

The *data source* option refers to a data source configured in Grafana. You can select any data source here, not just the one you're going to use the variable in. (This is especially useful when using lookup tables stored in a database or CMDB, but then using those results to query other systems for metrics or logs.)

Refresh determines how frequently the query will be run: *on dashboard load* will run the query once when the dashboard is opened, while *on time range change* will run the query every time a new time range is selected, either via the time selector at the top of the dashboard or when a user highlights a time range in a panel. On dashboard load is more efficient if your data never or rarely changes, but if you have values that can be added or removed frequently, like pod names in a Kubernetes cluster, using on time range change will make sure the right values are always available.

The *query* field is the actual query that will be run against your data source to return values for the variable to use. This needs to be a valid query for your data source. If it doesn't return any values, your variable won't have any either!

Regex lets you apply some filtering and alter the set of data by applying a regular expression. For example, if your query returns a set of long strings but you want to extract only a small substring value to use in your variable, applying a regular expression here will let you filter and edit down those values.

Finally, *sort* will sort the values before populating them into a dropdown list. You can choose to sort alphabetically or numerically either ascending or descending.

In our example, we can use this to retrieve the list of student names and IDs from the database with a query like

```
SELECT CONCAT(firstname, ' ', lastname) as __text, student_id as __value
FROM students;
```

Again, if you're not familiar with SQL, don't worry too much about the syntax here. What we're doing is retrieving the students' first and last names, concatenating them into one string, and pulling back their ID at the same time.

Tip   If you're using a data source that supports naming fields, like a SQL database, you can create key/value data from a query! Use __text to specify the key and __value to specify the value. This is a great trick to use for this sort of lookup – you can create a table that has a human-friendly name and a computer-friendly ID field and make beautiful, easy-to-use dashboards.

There are a couple of other fields you might have noticed here or on the custom variable type earlier. By default, Grafana will let users select only a single value for each variable. Enabling the *multi-value* option lets users select more than one value at a time, replacing the simple dropdown list with a set of checkboxes.

Turning on the *include all value* option will add a special option in your variable list for "All." You can set a special value for that variable here as well. In Figure 10-9, you'll see that this is set to '%'. This is the SQL *wildcard* character, meaning it will match anything. (It's also the reason we used the LIKE keyword instead of = in our query, as this tells SQL to match against wildcards as well as values.)

Putting this all together with a few more tweaks and variables lets us create a simple but usable dashboard like Figure 10-10.

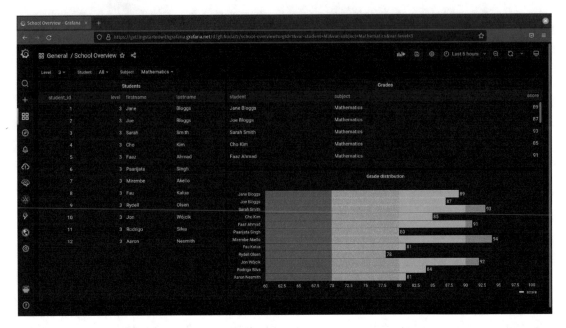

Figure 10-10. *Appropriate use of variables can let your users filter down to the data they need to see quickly*

Data Source Variables

Variables can be used for more than changing parts of queries. If you have more than one instance of the same type of data source, you can use variables to select which of your sources the panels in your dashboard will use for their queries.

Figure 10-11 shows the configuration for a data source variable. To start, set the *type* field to the data source that you will use. In this case, we're selecting between multiple different Loki data sources. You can optionally add a regular expression to filter the list of data sources that will be available to the variable; here, we've set one to show only sources that start with the string "Prod" to show only log sources for production environments, though you can use any regular expression here. The values that will be available are shown at the bottom of the page.

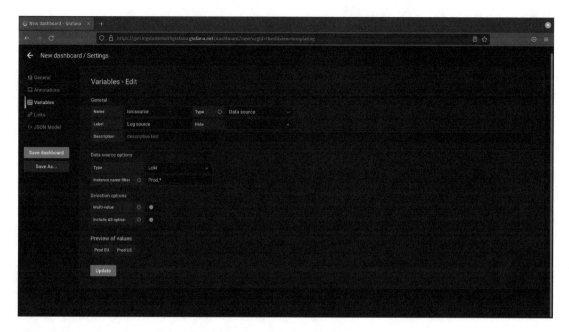

Figure 10-11. *Creating a data source variable*

Once you've created a data source variable, it will appear as a data source that can be used by panels on your dashboard, as shown in Figure 10-12. This will act as a special instance of the data source type that you've selected, and queries run using it as a source will be directed to the actual data source that is set in the variable. Changing the variable will immediately rerun those queries against the newly selected data source, and the dashboard will update automatically.

Figure 10-12. A variable data source in use. In this example, "${lokisource}" is replaced with the "Prod EU" Loki data source when the query is run

Chaining Variables

One of the most powerful – and least obvious – features of Grafana is that variables can be strung together, one after another. That is, the value of one variable can be used as input to a query in a second variable, and the value set in that second variable can be used as input to a third variable (and so on). This means that you can build logical flows of selecting values to visualize, letting a dashboard user narrow down or select the data that they want at each step until they get to the values they want without ever having to write a query.

Let's consider an example. Nearly every application and infrastructure component creates logs of some sort. So in most environments, people end up with a lot of logs to search through when something is going wrong. It's often the case that these logs are stored in different locations, and each source of logs might have its own set of values that can be used to filter and sort the logs.

If we consider the previous example, we can start with a data source variable to allow us to select which log store has the logs that we want to see, which helps. But we can extend the filtering and searching capabilities to build a powerful log searching dashboard for quick troubleshooting.

To start, let's create a second variable that lets us search through our logs for specific types of data. Labels are special metadata attached to log lines in Loki and used for fast searches, but the set of labels can be different in each Loki environment. So we can create a query variable to find the list of possible labels and have it run against the Loki data source specified by our data source variable. Figure 10-13 shows the configuration for this variable. Notice that the data source for this variable is set to ${lokisource} which means that the value of this variable will be used as the data source we search. The upshot of this is that when the data source is changed, this query will be run against the new data source, and the possible label values will be updated.

Figure 10-13. *Using a query variable to search a data source specified by a data source variable*

We could continue this pattern for as long as needed. You could use the value of this ${namespace} variable to run another query for values that are associated with the label, for example. As long as there are more values that can be useful in narrowing the search space down, it makes sense to keep adding variables.

Finally for this example, we will add a text box variable to let the dashboard user search for the specific text they want to find in their logs. As long as the query is constructed correctly using the variables we created, this search will look in only the specified data source and only at log lines tagged with the metadata they selected, as shown in Figure 10-14. This can be a great way to provide the power of the Grafana explore view inside of the context of a dashboard with more specific visualizations already configured.

Figure 10-14. *Multiple variables (top) are used in sequence in the panel query, letting the user find relevant data quickly without having to write their own query from scratch*

Repeating Panels with Variables

In addition to changing the data source or altering a query in a panel, variables can also be used to repeat a panel multiple times on a dashboard without having to copy and paste the panel. In Figure 10-15, you'll see an example of a panel that uses a variable in a query, much as we've seen earlier. The variable ${accesspoint} will be used in the Prometheus query in this panel to determine which WiFi access point's data to display. However, in this case, we've added a panel repeat option to the panel for this variable, highlighted on the right side of the image.

Figure 10-15. *A panel configured to use a variable value for repeating on the dashboard*

When this panel is saved, it will repeat for every value that is selected for the variable. (This means that in order to repeat, you will need to set the multi-value option on your variable!) Figure 10-16 shows the resulting dashboard if three WiFi access points are selected for this variable. While only one panel has been created, the repeat option tells Grafana to create an additional copy of the panel for each access point.

Figure 10-16. *While only one panel was created on this dashboard, the panel repeat option creates a new panel for each value of the selected variable*

Variables in URLs

If you've been experimenting with variables throughout this section, you might have noticed that variables and values started to show up in the URL bar of your browser. This is because Grafana uses the URL to track and store all the metadata about the dashboard you're currently using including the values of any variables. This has a lot of benefits, especially for working with Grafana programmatically. We'll look at more of those in Chapter 13 when we explore working with the Grafana API, but let's start here with managing variable values and linking to specific information in Grafana.

Passing Variables Between Dashboards

Generally, when using variables inside of Grafana, you don't need to think too much about setting or retrieving a value. As long as the variable has been created and has the same name on each dashboard, the selected values will just work. Creating a data link or panel link from one dashboard to another will pass along the value of all of the selected variables to the new dashboard, so you don't have to do anything extra.

Linking from External Tools

If you want to link to a Grafana dashboard with specific variables set from somewhere outside of Grafana, you can specify the variable values in the URL. All Grafana dashboard variables have the form var-name where *name* is the actual name of the variable as set in the dashboard settings. So, for example, to link to a Grafana dashboard with a variable called region and to pass in the value northamerica, you could create a URL like

```
https://yourname.grafana.net/d/glLNuda8a/regional-dashboard?orgId=1&var-
region=northamerica
```

The first part of the URL (everything up through orgId=1) you can copy and paste from your browser. Setting the value of variables is a matter of adding an ampersand (&) followed by var-name and the variable value. Grafana will parse this URL and do the rest for you.

Summary

In this chapter, you've seen how to work with variables to add dynamic features to your dashboards. You've learned how to add variables to a dashboard and reviewed some of the types of variables that are available. We looked at using variables as parts of queries as well as using variables to select a data source for a panel to run a query against.

We also reviewed how to use variables to repeat panels on a dashboard without having to copy and paste that panel multiple times, letting your dashboards adjust to the data that's available, even if it changes. Finally, we looked at working with variables inside of URLs and setting them both inside of Grafana and from external tools.

In the next chapter, we'll look at using Grafana to look for specific conditions and send alerts based on those.

CHAPTER 11

Alerting

Grafana is primarily a visual tool, designed to make it easy to present data to viewers in a clear and beautiful way. But it can also watch that data for you so that you don't have to constantly check to see what's happening in your environment.

The Grafana alerting system lets you set up rules that fire when certain conditions occur. For example, if the vibration frequency of a piece of machinery increases beyond a certain threshold, you might need to take some action: checking the equipment out, possibly even shutting it down. If this happens while someone is directly monitoring your dashboard, they might notice a spike in frequency, but if it's the middle of the night or the observer is distracted by other work, you still need to know that an issue is happening before your expensive equipment breaks completely!

Alerts can keep an eye on your data for you and take action when needed. In this chapter, we'll look at how to set up simple alerts, configure where those alerts are sent, and manage those alerts over time. This will serve as a solid introduction to Grafana's alerting capabilities, but there's much more available than we'll have space for, so be sure to review the alerting documentation for more details. The documentation is available at *https://grafana.com/docs/grafana/latest/alerting/unified-alerting/*.

Note In this chapter, we'll be using the unified alerting system that was introduced in Grafana 8.0. This new system became the default for new Grafana installs as of 8.3 and will become the default for Grafana Cloud as of the release of Grafana 9.0.

If you've just installed Grafana, you are all set! If you're using an older version of Grafana, you'll need to update and then enable the new alerting system. And if you're using an older version of Grafana Cloud, you may need to opt into the new system there as well. See `https://grafana.com/docs/grafana/latest/alerting/unified-alerting/opt-in/` for instructions on enabling the new alerting system for your environment.

© Ronald McCollam 2022
R. McCollam, *Getting Started with Grafana*, https://doi.org/10.1007/978-1-4842-8309-7_11

Alerting Configuration

The Grafana unified alerting system is managed from its own central configuration view. You can access this through the bell icon in the navigation bar on the left. Figure 11-1 shows the alerting configuration view with the navigation item highlighted. There are several tabs that are available in this view, each of which contains a set of information or configuration data. We'll look at these tabs individually.

Figure 11-1. *Grafana alerting is controlled centrally and uses tabs to divide functionality*

Alert Rules

The alert rules tab shows a list of alerts configured in your environment. If you're using Grafana Cloud, you'll see a number of alerts already predefined in your environment.

When an alert is active, that is, the alert condition that has been set is occurring, Grafana calls that a *firing* alert. Alerts can also be *normal*, meaning they are not firing, or *pending*, meaning that they are about to fire.

Note   It might not be immediately obvious what *pending* means for alerts; after all, we tend to think of alerts in terms of "this is broken" or "this is working." But sometimes it's okay to ignore brief bits of bad behavior. If a temperature reading goes above normal for a few seconds before going back down, it's probably not something worth getting concerned over – you might really only care when it stays too hot for several minutes.

Grafana can monitor data over time and alert when a condition has been met for a specified length of time to help prevent unnecessary alerts. When an alert condition has been met but it hasn't been long enough to actually trigger the alert yet, the alert becomes pending. If things go back to normal before the time period is up, the alert won't fire.

We'll look more at alert conditions later in the chapter.

By default, the alert rules view shows all alerts that have been defined, regardless of their state. You can filter this view to see only firing alerts by selecting the *firing* item under *state*.

There are other options for searching and grouping alerts here as well. Selecting a specific data source will show only alerts defined with data from that source. If you've added labels to your alert rules, you can search those as well. Finally, if you prefer to see alerts grouped by their state, for example, all alerts that are currently firing regardless of source, you can change the grouping method that Grafana uses.

Aside from searching and sorting, the alert rules tab lets you see the actual rules themselves, as shown in Figure 11-2. By clicking the triangle next to a rule name, you will get an expanded view with all of the details of the alert. The most important details in this view are the state, the expression, and matching instances. The *state* is whether the alert is normal, firing, or pending, as described earlier. The *expression* is the actual condition or value that is being checked, which we'll look into more later. *Matching instances* shows where the alert is currently firing. This last item is particularly important if you have an alert rule that is watching more than one source of metrics. For example, with a rule that checks the health of hundreds of servers, it's important to know which specific servers are unhealthy.

Figure 11-2. *Expanding an alert item reveals additional details*

It's not always obvious from an alert rule exactly what the state of a metric is or why it's causing an alert to fire. Clicking the *see graph* button will open the alert rule in the Grafana explore view, giving you a quick way to check the data visually.

Contact Points

The contact points tab contains the configuration for how alert messages are sent from Grafana and what those messages contain.

Message Templates

Message templates define what information is sent from Grafana when an alert is triggered. You can have multiple different message templates used for different groups of recipients, different environments or alert types, or different notification channels. For example, you might want more details in an email sent to your operations team than in a Slack message sent to management.

Message templates can embed variables that change the behavior and content of the message. In Listing 11-1, we show a snippet of a potential use of template variables, in this case to show a message depending on the number of firing alerts.

Listing 11-1. Using template variables in alert messages

```
{{ if gt (len .Alerts.Firing) 0 }}At least 1 alert firing{{ end }}
{{ if gt (len .Alerts.Firing) 10 }}More than 10 alerts firing!{{ end }}
```

This templating system allows programmatic control over messages and can build quite complex and flexible messages. However, for the most part, the built-in default templates will provide everything necessary for standard alerts.

Contact Points

Contact points define where alerts should be delivered. This can include email addresses, messaging channels like Slack or Discord, integrations with ticketing systems such as OpsGenie or VictorOps, and even webhooks for creating custom functionality.

You'll need to define at least one contact point in order to send alerts. To add a contact point, click the *new contact point* button in the contact points tab. You'll see a screen like Figure 11-3 where you'll be able to configure the channel that the notification uses. In this example, we'll set up a simple email. Other contact point types will require different information such as a URL to send messages to, credentials, or channel names that the message should be sent to.

Figure 11-3. *Configuring an email contact point requires only an email address. Other contact point types may have more settings*

You have the option in this screen of adding a message directly in the contact point. This can be configured the same way as message templates using the same variables and functions. If you leave this blank, the standard message template will be used.

It's a good idea to test the contact point once you've created it by clicking the *test* button. You don't want to wait until an alert is firing to find out that you mistyped part of the configuration! Figure 11-4 shows a test email sent from Grafana.

Tip   If you've deployed Grafana yourself, you'll need to make sure an email server is configured for your Grafana instance. To do this, look for the [smtp] section in your grafana.ini file. You can set up your mail server details there. If you're using Grafana Cloud, this is already taken care of for you, and you won't need to worry about it.

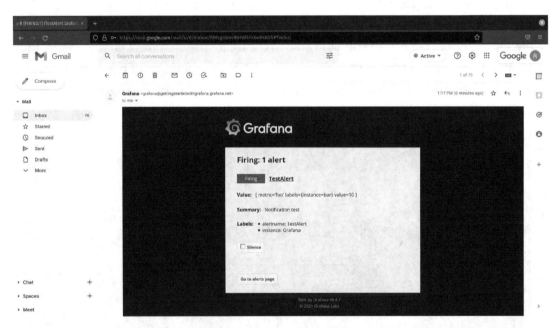

Figure 11-4. *A test alert from Grafana. In an actual alert, the variables in this message would be populated with real values*

You can also add multiple target systems for a single contact point. For example, you might want to have an email and a Slack message set simultaneously. To do this, just click the *new contact point type* button at the bottom of the page to add an additional contact point. (You'll need to configure this contact point as well for it to be used in alerts.)

Notification Policies

Notification policies define which alerts are sent to which contact points, as well as rules for when those alerts should and should not be sent. This is how your message templates are connected to specific groups of alerts and how you control when those alerts should be sent and when notifications should be muted.

Root Policy

The root policy is a "catch all" policy. Any alerts that don't have specific routing rules associated with them will be handled by this policy. Think of it as the default bucket that alerts are put into unless they have rules specifying otherwise.

By default, the root policy will be a predefined Grafana alert rule. This rule uses an example email address, so you will *definitely* want to either update that contact point to send to a valid notification channel or change the root policy to a contact point that you have defined. Figure 11-5 shows changing the root policy contact point to one that has been defined and tested.

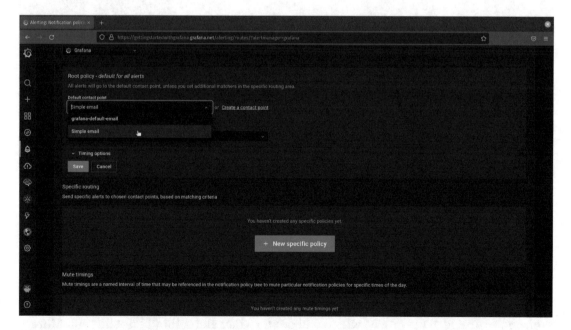

Figure 11-5. *It's important to set the root policy contact point to a valid channel – without this, you won't receive alerts without specific routing rules*

Specific Routing

While the root policy defines default behavior for alert notifications, it's likely that you will want to choose where and how to send notifications based on criteria like severity, environment, or time of day. Specific routings allow you to define rules for exactly this purpose.

You can create as many routing policies as you need to implement your workflow. Each routing policy allows you to filter alerts based on labels set on those alerts. If you include more than one label to match, *all* labels must match for the routing policy to apply. For example, in Figure 11-6, the routing policy will apply only to alerts that match `workload = production` and `datacenter = us_east`. An alert for `production` workloads in the `us_west` datacenter would not be matched by this rule.

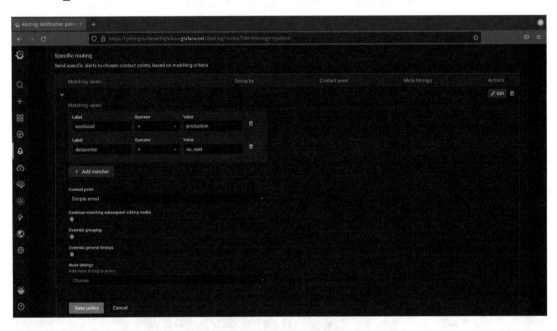

Figure 11-6. *Configuring a specific routing policy*

Specific routing policies also link to a contact point, so you can have different notification channels for alerts that match the labels you select. You can also optionally add a *mute timing* to the policy, which will define when alerts should not be sent. Mute timings are described as follows.

Mute Timings

Mute timings define periods during which alerts should not be sent. These are attached to specific routing policies, so you can have different routing policies take effect at different times of the day or month.

For example, you might have two shifts of workers who want to receive alerts. During the week, the main team should receive any notifications, but on the weekends a different set of employees should be alerted when something goes wrong. In this case, you could create a mute timing like the one in Figure 11-7 which mutes alerts only for weekends.

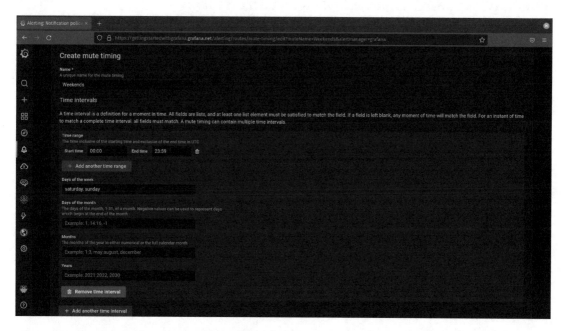

Figure 11-7. *A mute timing that prevents alerts from being sent on weekends*

Attaching this mute timing to a specific alert policy that contains the weekday employees notification policy will prevent them from being notified about alerts on Saturday or Sunday.

Caution   If you apply a timing like this, be sure that you also have another specific notification policy to deal with the alerts in another way! In this case, you'd also want to create a specific notification policy that sends alerts to the weekend team and attach a mute timing that suppresses alerts on weekdays. Without this step, you simply won't get alerts at all on weekends, which is probably not the goal!

If you have timings that are too complex to capture in a single rule, you can always click the *add another time interval* button to add more rules.

Silences

Silences are similar to mute timings in that they prevent alert notifications from being sent. However, where mute timings are used for ongoing, regularly scheduled muting, silences are intended as a one-time temporary tool.

Silences allow you to suppress alert notifications from being sent for a specific time, but do not reoccur. This is useful when you are performing maintenance for a specific time period or you have an ongoing incident that is being worked on and constantly receiving alerts is a distraction.

You can select labels to apply to, and like for muting rules, all labels must match for the silence to apply. As you'll see in Figure 11-8, you'll see a preview of the alerts that will be silenced when creating the rule. Be sure to check that you're not silencing alerts that should go through before saving the rule!

Figure 11-8. *Silences are a temporary, time-limited way to suppress alerts from firing*

Creating Alerts

While alerts can be created directly in the alert rules tab, the easiest way to create an alert is from a panel already populated with the data that you want to track. To create an alert, navigate to the panel edit view for the panel and select the *alert* tab, as shown in Figure 11-9. You can click the *create an alert rule from this panel* button to create an alert.

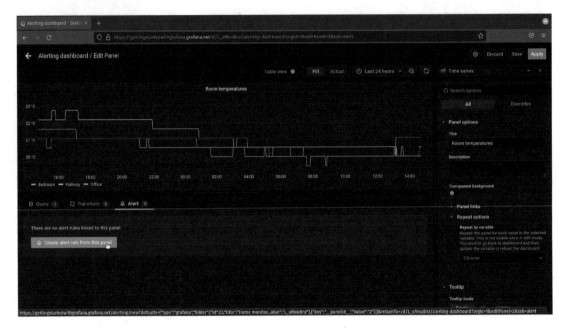

Figure 11-9. *The easiest way to create an alert is from a panel with the relevant data already populated*

When you create an alert rule from a panel, you'll be taken to the alert rule configuration page as shown in Figure 11-10. Here, you can adjust metadata about your alert and set the conditions for the alert to trigger. The alert configuration page is broken down into several sections, which we'll review as follows.

Figure 11-10. *The alert configuration page contains all of the information about an alert*

Rule Type

The rule type section defines some basic metadata for your alert. The rule name should be a brief, human-friendly description of the rule and will be what you see when browsing lists of rules in Grafana.

The rule type is a special field that lets you define where the rule will be evaluated. Grafana has support for using external alerting systems like Prometheus Alertmanager, which can be provided by a number of different projects. If you already have an Alertmanager instance configured that you want to keep using, you can select it here; otherwise, stick with the default Grafana managed alert.

Alert rules are contained in the same folder structure as dashboards. Your alert will default to the same folder that contains the dashboard you created it from, but if you want to change it to another folder, you can do that here.

Query to Be Alerted On

The query section is the most flexible part of alert configuration – and can be the most complicated as a result. You can define one or more queries against one or more data sources here in order to gather the data you'll need to evaluate the alert. Starting from a panel will automatically populate this section for you with the query (or queries) that makes up the panel, which makes things substantially easier.

When creating an alert, you'll see all the queries from your original panel as well as a new query at the end. This query is an *expression* rather than a direct query against a data source. This will define what condition you want to look for in your data to trigger the alert.

Grafana provides a simple query builder here to let you specify conditions. For example, in Figure 11-11, we've defined the alert condition for when the maximum detected temperature across all rooms in a house is above 20 degrees or the minimum temperature is below 5 degrees. You can add additional conditions to your query by clicking the + button below the list of conditions.

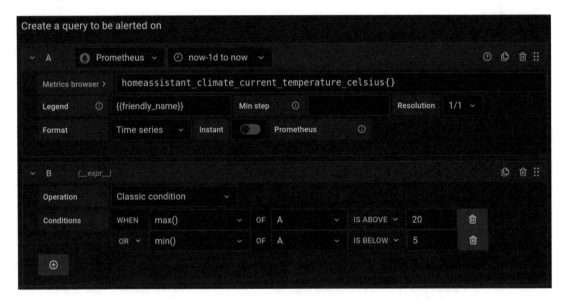

Figure 11-11. *An alert query that checks both upper and lower bounds of data*

Tip   It's important to think through the function you use when comparing a measurement against your threshold. Grafana defaults to using an average of all the metric series in your results, but in this case if one room were above 20 degrees and another room below 5, the average might be within the expected range and thus not trigger the alert. Look through the list of functions available and think carefully about how you're defining your alert query.

Alert Conditions

The alert conditions tell Grafana how to evaluate the alert query that you have created. Each alert condition has exactly one query that it looks at to evaluate the alert, which should be the condition that you defined earlier. That condition is then evaluated regularly to see if it is true or not. You can also tell Grafana how long it needs to remain true before an alert is firing, which is useful for things that might occasionally cross a threshold but come back out of the alert condition without too much trouble. In those cases, you usually want the alert condition to remain true for a while before actually triggering the alert.

You can optionally configure what to do if no data is received. If you stop getting data at all for this query, should that result in an automatic alert, an automatic healthy state, or to remain in the previous state until data is received?

Once you've defined all of these conditions, you should preview your alert to be sure it's in the state you expect. (It's often a good idea to set your threshold to a point that you know will trigger the alert right away, even if it's not the final threshold. Then you can adjust the threshold to the actual value and validate that the alert is no longer firing.) Figure 11-12 shows a configured alert condition that is being tested and actively firing.

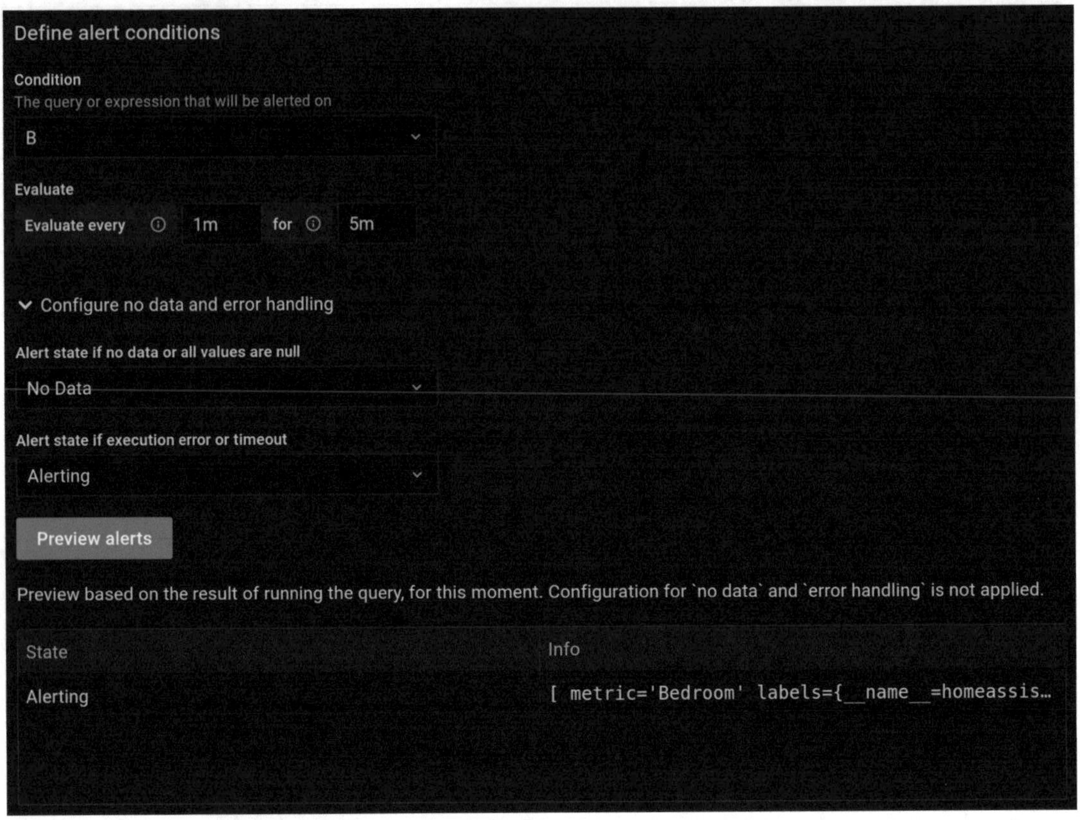

Figure 11-12. *Previewing an alert*

Tip   While it may be tempting to evaluate alerts every second, stop to consider whether you really need that level of granularity. Remember that every time an alert is evaluated, it will generate one or more queries against your data sources. Constantly running queries against your data sources may put an unnecessary load on them! It's best to evaluate your alert queries only as often as you really need to.

Alert Details

The final section in the alert configuration view is the alert details, shown in Figure 11-13. This is the information that will be sent along with your alert when it is triggered. If you've created an alert from a panel, the dashboard and panel will automatically be populated here so that a link to the dashboard can be generated in the alert. (If you want to change this to another dashboard instead, put its ID in the *dashboard UID* field.)

You can add other information such as a description of the alert or a link to a runbook containing information on how to troubleshoot the issue.

Finally, you can add labels to your alert. As we've seen earlier, these labels can be used for searching and sorting alerts as well as for applying notification policies and muting rules.

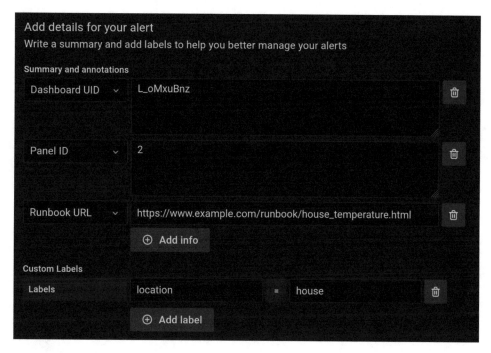

Figure 11-13. *Alert details provide additional context for your alert*

Summary

In this chapter, you've seen how to configure the Grafana alerting system, creating notification channels and routing rules to get your alerts to the right places. You've learned how to create muting rules to prevent alerts from being sent at the wrong times and silencing rules for briefly stopping alerts as needed. You've also learned how to configure alerts and create compound alert conditions to narrow in on specific criteria in your data.

In the next section, we'll start looking at more advanced Grafana usage. You'll see how Grafana can be managed at scale and be managed as a production service. You'll explore the Grafana API and see how to automate tasks and even full deployments. Finally, we'll look at the features provided by Grafana's commercial offering, Grafana Enterprise.

PART IV

Advanced Grafana

You've mastered the basics. You can stand up Grafana in a few minutes, and you're creating dashboards in your sleep. Now it's time to talk about making Grafana into something you can rely on completely: high availability, automated actions you control through your own scripts and applications, and enhanced security and reporting.

Part IV covers the parts of Grafana that you don't need to know for smaller environments and lightweight use, so depending on your requirements you may not need to spend much time here. But if observability is critical to your work, these topics will take your Grafana environment to full production readiness.

## CHAPTER 12

Advanced Deployment and Management

One of the advantages of Grafana is that it's self-contained and easy to set up. As you've seen already, you can download Grafana and be up and running on the platform and OS of your choice in just a few minutes. Sometimes, though, it's useful to add a bit more complexity to gain additional capabilities or integrate with other systems.

In this chapter, we'll look at ways to deploy Grafana to be more robust than the single self-contained instance you've installed so far. We'll explore ways of spreading Grafana's work across multiple servers for scaling and resiliency. You'll see how to refactor your environment so that the heaviest parts of alerting and reporting don't impact your main Grafana process. You'll also learn how to connect Grafana to other authentication systems, letting you manage logins from a directory service or a single sign-on (SSO) system.

External Grafana Databases

Note This section applies to Grafana environments that you have deployed and managed on your own. In Grafana Cloud, the Grafana database is managed for you as part of the service and cannot be changed.

It might have occurred to you to wonder where Grafana is storing all the data that it manages. Even though it's connecting to external data sources to retrieve the actual data that it displays, things like user account information, data source configuration, and dashboard layouts have to be stored somewhere locally.

© Ronald McCollam 2022
R. McCollam, *Getting Started with Grafana*, https://doi.org/10.1007/978-1-4842-8309-7_12

By default, Grafana uses a file-based database system called *SQLite*. SQLite is a mature (20+ years old) public domain SQL database engine that is designed to be small, fast, and reliable. Unlike most other SQL databases, however, SQLite doesn't run as a service that listens for queries and responds with results; instead, it runs as a library in memory within another program and performs all of its queries and writes on a file on disk.

Since SQLite is so small and doesn't require setting up any additional services, it's a perfect fit for most Grafana environments. But SQLite does have a few drawbacks: it's designed to be used by a single client at a time, and since it stores all data in a single file, it can be limited in size and speed as there's no good way to parallelize access to that one file. This means that your Grafana environment can grow too large for efficient use of SQLite, and if you want to have Grafana configured for high availability, you'll need to support multiple simultaneous systems accessing the data. Or you may just have requirements to use database systems that are already in use and being actively managed in your organization.

In any of these cases, you'll need to configure Grafana to use an external database service such as MySQL or PostgreSQL. Grafana's functionality is unchanged in this case; the only difference is the location where it keeps its own configuration data.

MySQL/MariaDB

To use MySQL (or MariaDB, a popular fork of MySQL), you'll first need to set up the database server itself. This server does not have to be on the same physical or virtual system as Grafana, but it does need to be reachable over the network from your Grafana instance.

You'll also need to create a database inside MySQL and a user account on the database for Grafana to use. When Grafana is configured to use the database, it will use this account to create and manage tables. And as Grafana is updated, tables may be altered, added, or dropped. As a result, the user account you create will need full privileges on the Grafana database. Listing 12-1 shows the commands used to create the database and user in MySQL. Note that these commands are run inside of MySQL, not at the command line of the server, so you'll need to connect to MySQL first.

Listing 12-1. Creating a MySQL database and user for Grafana

```
CREATE DATABASE grafana;
CREATE USER 'grafanauser'@'%' IDENTIFIED BY 'secure_password';
GRANT ALL PRIVILEGES ON grafana.* TO 'grafanauser'@'%';
FLUSH PRIVILEGES;
```

Running these commands will first create a database named *grafana*. It then creates a user named *grafanauser* who is allowed to log in from any host using the password *secure_password*. Finally, it grants full control of the new database to this user. You can change any of these values (and you should definitely set a real password!) but just be sure to be consistent when configuring the connection in Grafana.

Once you've set up your database and user account, you can add these to the Grafana configuration file, *grafana.ini*. (The location of your configuration file will vary depending on your installation; see Chapter 3 for more details.)

Inside of grafana.ini, look for the database configuration section, marked by [database] at the top. You'll see some example lines in the default Grafana configuration, but by default these start with a semicolon (;), meaning that they are comments. If you see any lines in this section that aren't comments, be sure to remove them or comment them out before adding your new configuration.

Listing 12-2 shows a database configuration section from *grafana.ini* that connects Grafana to a MySQL database. You need to be sure to use the exact values that you specified when creating your MySQL database and user earlier, as any typos here will prevent Grafana from starting up successfully. If your password contains the # or ; characters (which mark a line as a comment), you'll need to wrap the values in triple quotes, for example, """#my_more_secure_password;""".

Listing 12-2. You need to provide full connection details in grafana.ini to use a MySQL database in Grafana.

```
[database]
type = mysql
host = your_mysql_server:3306
name = grafana
user = grafanauser
pass = secure_password
```

Once you've made these changes, save the file and restart Grafana. If everything is configured correctly, you'll see your database quickly populate with Grafana tables. You will be able to log in after the initial database setup completes, which might take a few moments. If you don't see tables created or get an error when trying to connect to Grafana, double-check your configuration information in *grafana.ini* and review the Grafana logs for more information about what went wrong.

PostgreSQL

As with MySQL earlier, to use PostgreSQL as the backend data store for Grafana you'll first need to install and set up the PostgreSQL database itself. The database service doesn't need to be on the same physical or virtual server as Grafana, but it does need to be accessible over the network from your Grafana instance.

To start, you'll need to create a database for Grafana to use and configure a user account that has access to that database, as shown in Listing 12-3. These commands should be run inside of PostgreSQL, not on the command line of the Grafana server.

Listing 12-3. Creating a PostgreSQL database and user for Grafana

```
CREATE DATABASE grafana;
CREATE USER grafanauser WITH PASSWORD 'secure_password';
GRANT ALL PRIVILEGES ON DATABASE grafana TO grafanauser;
```

Running these commands first creates a database named *grafana*. It then creates a user account named *grafanauser* and gives it the password *secure_password*. Finally, it gives full control of the *grafana* database to the *grafanauser* account. You can change these values to whatever you like as long as you remain consistent when setting up the Grafana configuration. Be sure to set a real password here as well!

Once you've set up your database and user account, you can add these to the Grafana configuration file, *grafana.ini*. (The location of your configuration file will vary depending on your installation; see Chapter 3 for more details.)

Inside of grafana.ini, look for the database configuration section, marked by [database] at the top. You'll see some example lines in the default Grafana configuration, but by default these start with a semicolon (;), meaning that they are comments. If you see any lines in this section that aren't comments, be sure to remove them or comment them out before adding your new configuration.

Listing 12-4 shows a database configuration section from *grafana.ini* that connects Grafana to a PostgreSQL database. You need to be sure to use the exact values that you specified when creating your PostgreSQL database and user earlier, as any typos here will prevent Grafana from starting up successfully. If your password contains the # or ; characters (which mark a line as a comment), you'll need to wrap the values in triple quotes, for example, """#my_more_secure_password;""".

Listing 12-4. You need to provide full connection details in grafana.ini to use a PostgreSQL database in Grafana.

```
[database]
type = postgres
host = your_postgresql_server:5432
name = grafana
user = grafanauser
pass = secure_password
```

Once you've made these changes, save the file and restart Grafana. If everything is configured correctly, you'll see your database quickly populate with Grafana tables. You will be able to log in after the initial database setup completes, which might take a few moments. If you don't see tables created or get an error when trying to connect to Grafana, double-check your configuration information in *grafana.ini* and review the Grafana logs for more information about what went wrong.

High Availability Deployments

Note   This section applies to Grafana environments that you have deployed and managed on your own. In Grafana Cloud, high availability (HA) is managed and provided as part of the cloud service. You'll get HA there without any additional configuration required.

High availability in Grafana refers to having multiple Grafana instances running in the same environment with the same backend configuration database. In this configuration, one of your Grafana instances can go offline without affecting the availability of the full Grafana service.

The one prerequisite for HA Grafana is to use an external database for your Grafana backend, as outlined earlier. You must use MySQL/MariaDB or PostgreSQL for your Grafana environment to enable HA – the default SQLite environment will not work in an HA configuration, and **you risk data loss or corruption** if you try it!

Once you've configured a Grafana instance to use an external database, setting up HA in your environment is straightforward. You can deploy multiple copies of the Grafana service in separate servers, virtual machines, or containers and put a load balancer in front of them. As long as all of the Grafana deployments use the same configuration (by using identical *grafana.ini* files) and use the same backend database instance, Grafana will handle the rest. All user state is stored in the database, so there's nothing additional that you need to keep track of at the load balancer level.

Because you are using a load balancer to point at multiple instances, you'll need to tell Grafana what its canonical URL for the root of the environment is so that it can perform redirects to the user's browser as needed. The default of `localhost` will not work in an HA environment.

To do this, look in *grafana.ini* for the server configuration section which is headed with a `[server]` tag. You have two options here; you can set the values for `protocol`, `http_port`, and `domain`, or you can provide the full `root_url` directly. Listing 12-5 shows an example setting the first three options, and Listing 12-6 shows configuring Grafana with a single `root_url` value. Remember that if you see these values with a semicolon (`;`) in front of them, they are commented out – be sure to uncomment them when setting the values.

Listing 12-5. Setting variables to configure the Grafana root URL

```
[server]
protocol = http
http_port = 3000
domain = example.com
```

Listing 12-6. Setting the Grafana root URL directly

```
[server]
root_url = http://example.com:3000/
```

In both of these cases, the root URL for Grafana will be *http://example.com:3000/*.

Tip   Grafana is picky about the format of the root URL. The trailing slash (/) must be in place on the end of the URL for things to function correctly. Be sure to include it if you set the root URL directly!

Image Rendering Service

Note   This section applies to Grafana environments that you have deployed and managed on your own. In Grafana Cloud, image rendering is scaled and managed automatically as part of the cloud service.

Grafana is primarily a visual tool. So it makes sense that when sending information out of Grafana in an alert or a scheduled report, you want to be able to include a visual representation of the data as well. The Grafana image rendering system provides this functionality, acting as a way to make snapshot images of the more dynamic Grafana panels.

The easiest way to install and use the Grafana image rendering system is through a plugin available at *https://grafana.com/grafana/plugins/grafana-image-renderer/*. But the resource requirements for rendering images can be fairly high compared to the resources needed to run the Grafana service: it's recommended to provide at least 16 GB of RAM for just the renderer, compared to a fraction of that required for most Grafana deployments. If you have a highly available Grafana environment with many instances deployed inside it, this can quickly drive up the resource cost of running Grafana.

In these cases, it makes more sense to separate out the image rendering component into a separate service that each Grafana instance can use. You can set up one large instance (or several behind a load balancer for particularly large deployments) and have all of your Grafana instances send requests for image rendering to it.

Deploying the Rendering Service

The rendering service can be deployed either as a standalone Node.js application or as a Docker image provided by Grafana Labs. Either deployment type can be used by Grafana, and once the service is deployed, the connection to it is managed identically regardless of deployment type.

Docker Image

The Docker image is the simplest way to run a remote image renderer. The Docker image provided by Grafana Labs contains all the required components and prerequisites already installed and configured. The only thing you'll need is a working Docker environment.

To use the Docker container, you'll also need to expose the image rendering service's port, which by default is 8081. (You can change this to whatever port you like but remember to map it to port 8081 of the container.) For example, this command will download and start the image renderer, name it *image-renderer*, and expose the port:

```
docker run --name image-renderer --publish 8081:8081 grafana/grafana-image-
renderer:latest
```

The image will take a bit of time to download, but when this has finished, you'll see a success message letting you know the image rendering service is ready, similar to

```
{"level":"info","message":"HTTP Server started, listening at http://
localhost:8081"}
```

You can now set up your Grafana instance to use this rendering service as described in the following.

Standalone Node.js Application

In order to run the Grafana image rendering service as a standalone application, you'll need to have a fully working Node.js environment first, including npm and yarn. (If you don't have this already, please see *https://nodejs.org/* for more details.) You'll also need git installed to be able to retrieve the application source.

Start by downloading the image renderer source code:

```
git clone https://github.com/grafana/grafana-image-renderer
```

This will create a new directory, *grafana-image-renderer*, which we'll use to build the application. Once that's done, change to that directory and build the application as shown in Listing 12-7.

Listing 12-7. Building the Grafana image renderer from source

```
cd grafana-image-renderer
yarn install --pure-lockfile
yarn run build
```

The build may take a while to complete. When the build is finished, run the image renderer with the following command:

```
node build/app.js server --port=8081
```

This will start up your image rendering service on port 8081. (You can change this port to whatever you like; just remember to update any Grafana configurations that point to it.) You should see output like

```
{"level":"info","message":"HTTP Server started, listening at http://
localhost:8081"}
```

This means that your image renderer is running and ready to use. See the next section on configuring Grafana to make use of it!

Configuring Grafana to Use the Remote Renderer

Verifying Your Renderer

Before setting up the external rendering service in Grafana, it's a good idea to verify that it's up and running first. To test this, you can load a special URL on your server that will return the service's version: */render/version*.

You'll need to know the server and port that your service is running on. Once you have that, you can use a web browser or a utility like curl to check this URL. For example, if your rendering service is on *renderer.example.com* running on port 8081, you can request *http://renderer.example.com:8081/render/version* in your browser or curl. If everything is working, you should see a response with the version number, similar to

```
{"version":"3.4.1"}
```

If you see output like this, then your image renderer is ready to go!

Configuring Grafana

Setting up the external rendering service connection requires telling Grafana where it can find the remote service. To do this, edit *grafana.ini* and look for the rendering configuration section, which is headed with a [rendering] tag.

There are several options that can be set, but the two that are required are *server_url* and *callback_url*. The first, server_url, tells Grafana where to find the rendering service. The second, callback_url, contains information about where the rendered image should be sent – without this, the rendering service wouldn't know what to do with its output after it finishes rendering an image.

Listing 12-8 shows an example configuration for the rendering section of *grafana.ini*. In this example, the rendering service is running on port 8081 (the default) of a server called *renderer.example.com*. Grafana itself is running on port 80 on *grafana.example. com* and is set in the callback URL. Note that the actual rendering service listens on a specific path, */render*. You'll need to include that in your URL when configuring a remote rendering service.

Listing 12-8. Configuring a remote rendering service requires both the remote URL and the local Grafana URL.

```
[rendering]
server_url = http://renderer.example.com:8081/render
callback_url = http://grafana.example.com/
```

Once you've made this change, restart Grafana. You should see a line like this in the Grafana logs:

```
INFO[03-13|16:07:00] Backend rendering via external http server
logger=rendering renderer=http version=3.4.1
```

This indicates that Grafana has successfully connected to the external renderer. If you don't see this line or see an error, check your configuration and ensure that your rendering service is running.

Backup and Restore

Because Grafana exposes its configuration through APIs and JSON formatted data, it's designed to allow you to manage its configuration as code. And in fact, for most cases, this is the best way to go – using the APIs and native configuration data lets you keep dashboards and data sources as code in source code repositories, tracking changes as they occur and letting you roll back specific components or dashboards without affecting the entire environment.

That said, it can also be useful to have a full snapshot of an environment for backup or disaster recovery purposes. There are only a few items that need to be copied to enable you to completely restore a Grafana environment: any installed plugins, the Grafana configuration database, and your Grafana configuration file. Everything else is part of the Grafana deployment package so you can back up only the essentials.

Backing Up

To start, make a copy of any installed plugins in your Grafana environment. This can be done by copying the *plugins* directory inside of your Grafana data directory. (The data directory can vary depending on your installation; see Chapter 3 for more details about this.) Copy the full contents of this directory, including subdirectories and all files.

Next, make a copy of your *grafana.ini* file. Like with the location of the plugins directory, the location of this file varies depending on how you have installed Grafana. Chapter 3 will help you find it if you've forgotten where it is.

Finally, back up the Grafana database itself. This will vary a bit depending on whether you're using the default SQLite deployment or an external database.

SQLite (Default)

If you haven't configured an external database for your Grafana environment, you can simply copy the SQLite database file. This should be in your Grafana data directory and be named *grafana.db*. This contains all of the configuration and dashboard data for your Grafana deployment.

MySQL

If you've configured an external MySQL database for Grafana, you'll need to export all of the tables for the Grafana database to a file. This can be done in the MySQL GUI (or any number of other MySQL frontends) or from the MySQL command-line tool. For example, if you have a database named *grafana* that is accessed with the username *grafanauser* and password *secure_password*, then you can back up your Grafana instance with a command like

```
mysqldump -u grafanauser -psecure_password grafana > backup.sql
```

This will create a file called *backup.sql* that contains all of your Grafana configuration data.

Tip   Note that there's no space between the -p option and the password when using `mysqldump`. If you get an authentication error when connecting, be sure that you don't have an extra space in this command!

PostgreSQL

For a Grafana instance that is configured to use an external PostgreSQL, you'll need to back up the contents of the database to a file so that it can be stored offline. Similarly to MySQL, you can do this with a GUI tool such as *pgAdmin* (or any number of similar tools). This can also be done from the command line with the built-in PostgreSQL tools. For example, if you have a database named *grafana* that is accessed with the username *grafanauser*, then you can back up your Grafana instance with a command like

```
pg_dump -d grafana -U grafanauser > backup.sql
```

This will prompt you for a password for the *grafanauser* account. Once you put in this password, pg_dump will create a file called *backup.sql* that contains all of the contents of your *grafana* database.

Restoring

Restoring Grafana from a backup is simply a matter of replacing the copied files and configuration data in a newly deployed Grafana environment.

To start, install Grafana normally. (Refer to Chapter 3 for more details on installing Grafana.) Once Grafana is deployed, be certain that it is not running – shut down the service or terminate any running Grafana processes. Then you're ready to restore your backup.

Caution   Be sure to deploy the same version of Grafana that you took your backup from! Code changes and database schema changes can happen between even minor releases of Grafana, meaning that restoring a backup configuration over the top of a different Grafana release may not work or might lead to compatibility issues down the line. Always restore with the same version that you backed up and then upgrade as desired.

With your new Grafana deployment, start by copying the *grafana.ini* file back to the proper location for the new installation. It's a good idea to review the contents of this file to make sure that nothing in the environment has changed since the backup occurred.

Next, copy back in the *plugins* directory that you backed up before. Be sure that all subdirectories and files are copied in this step; if parts are missing, you'll wind up with broken plugins.

Caution   You also need to check file and directory permissions carefully here. Whatever user account is running, the Grafana service will need to be able to read and write these directories and files. If the plugins directory is read-only for the Grafana service, you'll be unable to install new plugins and may have unexpected behavior from existing ones.

Finally, you'll need to restore the configuration data that you backed up previously. As before, the way this works varies a bit depending on if you're using the default SQLite deployment or if you have configured an external database.

SQLite (Default)

To restore a default SQLite installation of Grafana, simply copy the backed-up *grafana. db* file to the Grafana data directory. Carefully check the permissions when you do so – the user or service account that Grafana is running as **must** have full read and write permissions to this file! If these permissions are missing, your Grafana environment will not function correctly.

Once you've copied this file in, start the Grafana service and your backup should be running.

MySQL

To restore a Grafana environment with an external MySQL database, you'll first need to restore the full database backup into your MySQL environment. Like with backing up, this can be done through the MySQL GUI or various other UIs or via the command line. To restore a MySQL environment from a file named *backup.sql* into a database named *grafana* that has been set up with a user account of *grafanauser* and a password of *secure_password*, you would use the following command:

```
mysql -u grafanauser -psecure_password grafana < backup.sql
```

This will run the SQL statements in the *backup.sql* file to recreate the Grafana table structure and re-add your data. Once this is finished, start the Grafana service and verify that your data is accessible.

> **Tip**   As with the backup instructions earlier, note that there is no space between the `-p` option and the password in the `mysql` command. You'll also need to be sure to create the database and user credentials in the database for this restore to work. See the preceding section on configuring an external MySQL database for Grafana for more information on creating user accounts in the database.

PostgreSQL

To restore a Grafana environment configured to use an external PostgreSQL database, you'll need to first restore the database structure and contents from your backup. This can be done using *pgAdmin* or other GUI tools or from the command line. For example, to restore a backup from a file named backup.sql to a PostgreSQL environment containing a database named grafana with a user account of grafanauser, you would use a command line like

```
psql -U grafanauser -d grafana < backup.sql
```

This command will prompt you for a password for the *grafanauser* account. Once you input it, the *backup.sql* file will be read and executed, recreating the structure and data from your PostgreSQL environment. After this completes, start your Grafana service and verify that your data has been restored.

Tip   Much as with MySQL earlier, you need to be sure that the account information has been properly configured for your PostgreSQL database before running this command. See the preceding section on configuring an external PostgreSQL database for more information on creating user accounts in PostgreSQL.

Advanced Authentication

We've looked at how to manage Grafana logins in previous chapters, both as credentials stored in the Grafana database and managed by Grafana and as credentials managed by Grafana Cloud.

You can also connect Grafana to other directory systems such as Active Directory or LDAP and to third-party authentication systems like OAuth, a single sign-on (SSO) system commonly used for authenticating on websites with your email provider without requiring you to have separate accounts for each service.

Authentication and SSO systems are an expansive and nuanced set of topics, and it would be impossible to cover every possible configuration here. We'll explore some example setups and look at where in Grafana these settings are configured, but when setting up your own authentication integrations, be sure to consult both the directory or authentication system's documentation and the Grafana authentication documentation at *https://grafana.com/docs/grafana/latest/auth/*.

Finally, be sure to see Chapter 14 to see some additional authentication options that are available with Grafana Enterprise.

Note   The configurations explored as follows assume you're working with a local Grafana instance. These authentication options are also available in Grafana Cloud. In order to set up these integrations in Grafana Cloud, open a support ticket in your cloud environment and the Grafana Labs team can help you connect.

LDAP and Active Directory

Enabling LDAP authentication in your Grafana environment allows users to connect to Grafana using their user credentials that are stored and managed in a centralized directory service. This includes Microsoft's Active Directory service as well as other LDAP servers such as OpenLDAP or Apache Directory Studio.

Using LDAP authentication does not automatically disable Grafana-managed logins, so you can have a mix of local and directory-based users. This is especially useful for configuring local administrator accounts that are not necessarily managed at the directory level.

Groups of users in the directory can be mapped to specific roles in Grafana. For example, members of a group called "VIPs" in your directory can be granted admin permissions in Grafana by default, while "managers" or "directors" might be editors. Permissions are additive, so a member of both the "VIPs" and "managers" groups will have full admin access. (More granular access permissions and team mappings are part of Grafana Enterprise and are covered in Chapter 14.)

Let's walk through an example of configuring LDAP access for a Grafana server. To do this, we'll use the example LDAP environment provided by Forum Systems as described at *www.forumsys.com/2014/02/22/online-ldap-test-server/*. This is a read-only LDAP environment provided publicly for testing, so it's a great way to see how directory integration in Grafana works without having to set up your own LDAP or Active Directory environment.

Enabling LDAP

To enable LDAP login support in Grafana, you'll need to update your Grafana configuration in *grafana.ini*. Look for the LDAP section marked by [auth.ldap]. The settings in this section let you enable or disable LDAP authentication in Grafana globally as well as tell Grafana where to find the specific settings associated with your environment. Listing 12-9 shows an example of this section, enabling LDAP and pointing at a configuration file that will contain authentication information.

Listing 12-9. Enabling LDAP in Grafana

```
[auth.ldap]
enabled = true
config_file = conf/ldap.toml
allow_sign_up = true
```

Setting *enabled* to true turns LDAP support on in your Grafana environment. The next value, *config_file*, tells Grafana where to look for the connection and user information for your directory service. In this example, we're calling this file *ldap.toml* and keeping it in the Grafana *config* directory, but this can be any file that is readable by the account running the Grafana service. Finally, *allow_sign_up* will determine if users that log in for the first time with LDAP credentials are allowed. If this is set to *true*, anyone with valid credentials in the directory will be able to log in to Grafana. If it's set to *false*, then only users that have previously logged in (or been created in Grafana locally with the same username) will be able to log in. No new users will be added automatically in this case.

Configuring LDAP

Once LDAP is enabled, you need to tell Grafana how to connect to your directory service and where to look for valid user accounts. This information is stored in the file provided to the *config_file* setting in *grafana.ini*. Listing 12-10 shows an example of this file with some comments included for clarity. We'll break this down section by section.

Listing 12-10. A complete LDAP configuration file for Grafana contains both connection details and user mapping information.

```
[[servers]]
# LDAP server(s) connection info:
host = "ldap.forumsys.com"
port = 389
# No encryption required here; see docs for LDAP SSL settings
use_ssl = false
```

```
start_tls = false
ssl_skip_verify = false

# How Grafana authenticates to the LDAP server:
bind_dn = "cn=read-only-admin,dc=example,dc=com"
bind_password = 'password'

# The LDAP search string to use to find valid users:
search_filter = "(uid=%s)"

# Where in the directory to find those users:
search_base_dns = ["dc=example,dc=com"]
group_search_filter = "(&(objectClass=groupOfUniqueNames)(uniqueMember=uid=
%s,dc=example,dc=com))"
group_search_base_dns = ["dc=example,dc=com"]
group_search_filter_user_attribute = "uid"

# What fields to map in the directory to Grafana user info:
[servers.attributes]
name = "givenName"
surname = "sn"
username = "uid"
member_of = "DN"
email =  "mail"

### Below are mappings of LDAP groups to Grafana roles

# Make members of the 'chemists' group admins
[[servers.group_mappings]]
group_dn = "ou=chemists,dc=example,dc=com"
org_role = "Admin"

# Make members of the 'scientists' group editors
[[servers.group_mappings]]
group_dn = "ou=scientists,dc=example,dc=com"
org_role = "Editor"
```

```
# Make everyone else viewers
[[servers.group_mappings]]
group_dn = "*"
org_role = "Viewer"
```

Let's look at this in more detail. The first section, everything under the [[servers]] heading, defines how Grafana connects to the LDAP server. (You can have multiple servers listed here by separating them with spaces. In this example, we'll keep it as simple as possible.)

The *host* and *port* settings tell Grafana where the LDAP service is, in this case at *ldap. forumsys.com* on port *389*. We also turn off all the various LDAP encryption options, telling Grafana not to use SSL for this connection as it's not required here. In production systems, you'll likely need to turn SSL on and encrypt all connections. In these cases, you'll need to provide a path to a public key to encrypt your requests; see the Grafana LDAP configuration documentation for an example of this.

Next, we provide credentials to connect to the LDAP service. The *bind_dn* is effectively a username, and *bind_password* is its associated password. These settings will vary for other environments, but in this case are provided as public credentials by Forum Systems.

The *search_filter* setting specifies how to find users. LDAP directories can contain much more than just user account information, so this tells Grafana the query to use to find only user information. Again, this will vary from environment to environment.

The next group of settings further refine this search filter, specifying where to look for users and groups within the directory. In this case, Forum Systems is providing a directory of users in the *example.com* domain, which is then formatted for LDAP searches as *["dc=example,dc=com"]*. These values will differ in every environment and will need to be provided by your directory administrator.

The final settings in this section of the file control how attributes in the directory are mapped to their corresponding fields in Grafana. For example, Grafana has a *name* value which in this directory corresponds to a field named *givenName*. Similarly, what Grafana calls *email* is named *mail* in this LDAP server. This section tells Grafana which fields to use internally when authenticating users.

After the server configuration section, we have several sections that tell Grafana how to map users in the directory to the appropriate permissions in Grafana. You can have as many of these sections as makes sense for your environment, but for simplicity here we have just three. Let's look at one of these in more detail:

```
[[servers.group_mappings]]
group_dn = "ou=chemists,dc=example,dc=com"
org_role = "Admin"
```

The first line tells Grafana that this section defines a mapping of a group in the directory to a role in Grafana. As stated earlier, there can be as many of these sections as needed for your environment.

When configuring this section, *group_dn* specifies the name of a group in the LDAP directory. Any user that is a member of this group will receive the specified role. In this case, anyone who is a member of the *chemists* group in the *example.com* domain will be mapped to the value in the *org_role* setting.

In this case, any members of the *chemists* group in the directory will be granted the *Admin* role in Grafana.

Once you've configured these settings, restart Grafana. If everything is set up correctly, you'll be able to log in to your Grafana instance with users from the LDAP environment. Some examples to try are *curie* for a member of the *chemists* group, *euler* for a *mathematicians* member, and *einstein* as a member of *scientists*. (The password is *password* for all accounts.)

Tip   There are a lot of moving parts in this configuration, and it can sometimes be hard to figure out what's going wrong when it doesn't work. You can get a lot of great debugging information by adding this to the top of your LDAP configuration file (*ldap.toml* in the preceding examples):

```
[log]
```

```
filters = ldap:debug
```

This will cause Grafana to add LDAP debugging information to the Grafana log file. Just be sure to turn it off again once you've gotten things working, as debug logs can be very high volume and waste filesystem space.

Using LDAP

Once LDAP is configured in Grafana, there are a few immediate changes. As we've seen earlier, you can log in as users that are specified in the directory service.

But there's also a new addition for Grafana Administrators (such as the default *admin* account): an LDAP mapping explorer. This new view lets you test what the permissions of LDAP users will be without having to log in as them individually.

The LDAP tab is found in the Server Admin panel. Submitting a username here will show you where the user is validated (if at all), what information is returned from the LDAP server about them, and what their role will be in the Grafana environment. Figure 12-1 shows an example of this with the LDAP environment described earlier.

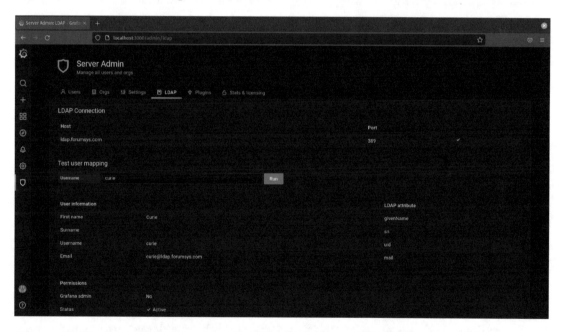

Figure 12-1. *Just when you thought this chapter wouldn't have any pictures, the LDAP server admin panel appears! It shows values mapped into Grafana from a remote directory server*

OAuth

OAuth is an open authorization framework that allows one website or application to pass authentication information to another. It's frequently used to log in to websites using credentials from Facebook, GitHub, LinkedIn, email providers, and many other service providers.

Grafana supports OAuth2 as a login mechanism, meaning that you can allow users to connect to your Grafana environment by using login information provided by one of the preceding providers or through a single sign-on (SSO) system that is managed centrally in your organization.

Much like with LDAP, there are a number of different providers and configuration options that are possible. The full documentation on Grafana OAuth configuration is available at *https://grafana.com/docs/grafana/latest/auth/generic-oauth/* and provides much more in-depth coverage of this topic. In this section, we'll take a look at configuring a single, commonly used OAuth provider – in this case, GitHub – in Grafana. This gives us a standard framework to explain how OAuth configuration works, but you'll need to make some changes to adapt this to your specific environment.

Creating an OAuth Application

In this example, we'll be using GitHub as our OAuth provider. GitHub provides free accounts and the ability to easily create an OAuth application that can be used to log into Grafana. If you don't have a GitHub account already, you can sign up for one for free at *https://github.com*.

To start, create an OAuth application on GitHub. The instructions for this are available at *https://docs.github.com/en/developers/apps/building-oauth-apps/creating-an-oauth-app*.

When creating your application, you'll need to provide two key pieces of information: the application's *homepage URL* and a *callback URL*. The homepage URL will be simply the URL for your Grafana instance. For example, if you are hosting your Grafana instance on *grafana.example.com* on port 3000, your homepage URL will be *http://grafana.example.com:3000/*.

The callback URL is a special path that tells the OAuth provider where to send authentication data when a user has successfully connected. There's a predefined target for this in Grafana called */login/github*. So using our preceding example, the callback URL would be *http://grafana.example.com:3000/login/github*. Use your own URLs when configuring the GitHub OAuth application.

Note   It's worth talking about the Grafana *root_url* setting and the special localhost URL that Grafana defaults to here. If you're running Grafana locally on your desktop or laptop on the default port, you can use `http://localhost:3000/` as your homepage URL and `http://localhost:3000/login/github` as the callback URL when creating your GitHub application. This will work as long as you are logging in only from the same machine that is running Grafana.

This is fine for testing and a great way to see how OAuth configuration works in Grafana. But in production, you'll need to put a "real" URL in here. Without that, your login will fail from any system other than the one running Grafana. You need to be sure that "real" URL matches what is set in the *root_url* setting in *grafana.ini*. Be sure to check this and set its value to the URL that you want people to use to connect to your Grafana server.

After you've set the GitHub application up, make a note of the *client ID* and the *client secret* that are generated. You'll need these to configure Grafana to talk to your application.

Configuring OAuth in Grafana

Authentication settings in Grafana are controlled in the Grafana configuration file, *grafana.ini*. To enable your OAuth application, search for the GitHub authentication section that starts with [auth.github]. Listing 12-11 shows an example configuration for a GitHub OAuth application.

Listing 12-11. A GitHub OAuth application configured in Grafana

```
[auth.github]
enabled = true
allow_sign_up = true
client_id = 32XXXXXXXXXXXXXXXXXX
client_secret = 1cceXXXXXXXXXXXXXXXXXXXXXXXXXXb2ad
scopes = user:email,read:org
auth_url = https://github.com/login/oauth/authorize
token_url = https://github.com/login/oauth/access_token
```

```
api_url = https://api.github.com/user
allowed_domains =
team_ids =
allowed_organizations =
```

The first line, *[auth.github]*, defines a Grafana authentication configuration for GitHub, and setting *enabled* to *true* in the second line turns this integration on. Everything below this describes the integration. Let's look at this in more detail.

The *allow_sign_up* setting determines whether users that have not previously logged in can be added by this integration. If it's *true*, then anyone who authenticates through OAuth and meets the requirements outlined in the rest of the section will be able to log in. If it's *false*, then only existing users can log in, and new users will be rejected.

The *client_id* and *client_secret* are the values that are created in GitHub as part of your application. These will be used by Grafana to connect to GitHub and have it properly match users to the Grafana environment for login.

Next, the *scopes* setting determines what information Grafana will receive from the OAuth provider. In this case, it asks for user and email information as well as what organizations in GitHub that user belongs to.

The three URLs, *auth_url*, *token_url*, and *api_url*, are resources provided by the OAuth provider, in this case GitHub. These are part of the OAuth system itself and will be provided by your OAuth provider. If you're using GitHub as a provider, these will already be populated, and you can keep the defaults.

The final three lines let you limit which GitHub accounts can authenticate to your Grafana environment. If you set *allowed_domains* to *example.com*, then only users that have email addresses from example.com domains can log in. Similarly, *team_ids* and *allowed_organizations* let you specify the GitHub teams and organizations that will be given access to Grafana. If left blank, as in this example, anyone with a valid GitHub account will be able to log in.

Once you've configured OAuth, restart Grafana. When you visit the login page, you should see a new option for GitHub logins as shown in Figure 12-2.

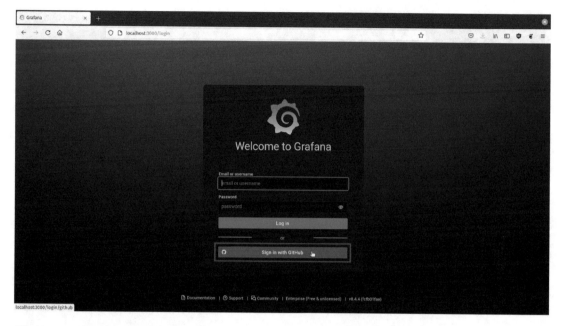

Figure 12-2. *Configuring GitHub as an OAuth provider gives a new sign-in option on the Grafana login screen*

Using OAuth

The first time someone logs in with OAuth in your environment, they'll be sent through a new login flow. They'll log in to the OAuth provider, in this case GitHub, and be asked if they want to provide their account data to the target application, in this case Grafana.

Figure 12-3 shows an example of this flow using GitHub as the OAuth provider. Once the user accepts, they'll be logged in to Grafana and able to use the environment.

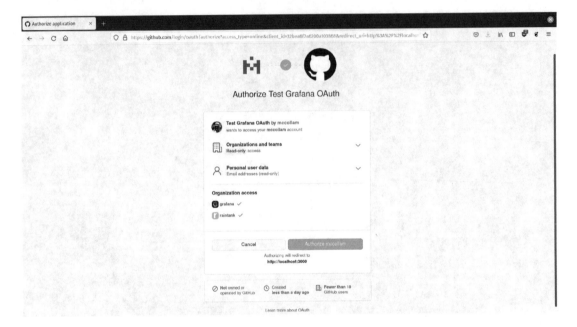

Figure 12-3. *Authorizing a test Grafana application in GitHub OAuth*

Summary

In this chapter, we reviewed more advanced ways to deploy and manage Grafana. You've seen how to connect Grafana to an external database, increasing scalability and providing pathways to run Grafana in a highly available fashion. You learned how to configure an external image rendering service for your environment, further expanding scalability and ensuring that the Grafana instances that are being accessed by end users are not slowed down by reporting and alerting. We looked into methods for backing up and restoring Grafana configurations. Finally, we explored several options for allowing external authentication methods for Grafana.

In the next chapter, we'll look at more automated ways to work with Grafana. You'll see how to leverage the Grafana API to automate common tasks and integrate with other applications. We'll also look at the Grafana provisioning system which allows you to treat Grafana configuration as code and quickly duplicate an entire Grafana environment.

CHAPTER 13

Programmatic Grafana

Up until now, all the ways that we have seen to use Grafana have been through its web user interface. This UI is fantastic for creating dashboards, laying out panels, and organizing resources. But when you need to manage Grafana at a large scale or as part of a continuous deployment system, having to click buttons and type information into text boxes can be slow and limiting. Fortunately, Grafana provides a rich *application programming interface* (API) for automating these actions.

In this chapter, we'll explore some of the ways you can use the Grafana API to automate tasks. You'll see how to configure API access to your Grafana environment, how to export and import dashboard and data source configurations, and even how to check on the status of your Grafana environment programmatically. Using these APIs, you'll be able to integrate Grafana into a continuous deployment model and to treat dashboards and configuration as code, backing them up and making them available for storage in a source code version control system.

We'll also take a look at the Grafana provisioning system, which lets you deploy and update dashboards and data sources as part of a Grafana deployment without calling the APIs directly, using configuration files instead. Provisioning lets you package up a full Grafana environment as a ready-to-run system without having to script out API calls at all.

REST APIs

Before diving into the Grafana API itself, it's worth backing up a bit and understanding the principles that the Grafana API is built upon. Like most modern web applications, Grafana is based on the idea of *representational state transfer*, or REST.

© Ronald McCollam 2022
R. McCollam, *Getting Started with Grafana*, https://doi.org/10.1007/978-1-4842-8309-7_13

> **Note**   If you're already familiar with using REST APIs generally, it's safe to skip ahead to the next section. If not, don't worry – while there's a lot that can be said about REST APIs, we'll keep this nice and light!
>
> If you *really* want to get into the full computer science underpinnings of REST, you can check out the doctoral thesis of Roy Fielding who formalized the concept. It's available online at *www.ics.uci.edu/~fielding/pubs/dissertation/ top.htm.* But for our purposes, you really just need to know the basics of sending requests and receiving responses from a web application.

REST is really just a fancy (and formal) way of saying that every request that is sent to the application is independent of any other request. That is, when you access a URL and request a resource, the web application doesn't automatically remember who you are or what requests you've made in the past. You need to provide this context about your request, which REST calls *state*, every time you make a new request.

The upshot of this is that each time you ask for a resource from or send information to Grafana, you need to include all the state that Grafana needs to fulfill that request. This includes things like authentication, so you'll need to include your credentials with every request. It also means that you have to be explicit about what resource you are asking Grafana to modify – just because you made an update to a specific dashboard a moment ago doesn't mean that Grafana will remember that! You'll need to explicitly tell Grafana everything about what you want it to do, where you want that to happen, and who you are with each request.

(You might ask why REST is so popular if it's so limiting. There are a number of reasons for this, but a big one is that it makes a system to handle these types of requests simpler to build, test, and scale. As a REST application, you can simply add new servers to a Grafana environment, and it will immediately be able to handle more simultaneous users. Non-REST systems are far more difficult to scale, if they can at all.)

Using a REST API Client

Most uses of the Grafana API will be within programs or scripts that you manage for the purposes of automating your environment. However, when learning and experimenting with an API, it can be useful to use an interactive REST API client.

There are a vast number of REST API clients available, any of which should work with Grafana. For the purposes of illustration in this chapter, we'll use Postman, one of the most popular clients in use at the time of the writing of this book. Postman has a number of advantages: it's cross-platform and available for Linux, MacOS X, and Windows; it has a web version that is accessible from anywhere with a browser; and while Postman offers paid services, the client itself is free and open source. If you already have an API client that you are comfortable with, please feel free to use it instead, as there will be nothing specific to Postman in this section.

Postman is available for download from *www.postman.com/downloads/*.

Managing Grafana API Keys

API keys are similar to user accounts. They have a name and a role which determines their level of access to your environment. Unlike user accounts, however, they can't be used to log in to Grafana interactively – they grant access only through the API itself, not the web user interface. Additionally, API keys have a *time to live*, a period of time for which they are valid. After this time runs out, the API key will no longer work.

Only users with the admin role can manage API keys in Grafana. When logged in as an admin user, navigate to the API keys section of the Configuration view. Figure 13-1 shows the API key management view with the link to this view in the configuration menu highlighted. At the top of this view is a search box, letting you find existing keys easily in a long list of keys. Below this is a list of existing keys showing the key name, its role, and expiration date (if any). By default, only currently valid keys will be displayed. To show expired keys, activate the "include expired keys" option.

Figure 13-1. *The API keys section of the Grafana configuration view shows a list of existing API keys*

Note   If you're using Grafana Cloud, there's also a section in the account management page that allows managing API keys. These keys apply to more than just Grafana itself and include the ability to define keys for publishing metrics and logs to the additional services Grafana Cloud provides for these. We'll be looking specifically at Grafana API keys and managing them directly in the Grafana admin interface.

Adding API Keys

To add an API key to Grafana, click the "Add API key" button to the right of the search box. This will open a new set of options as shown in Figure 13-2. The name field is for a human-readable identifier for your API key. Try to make this clear and descriptive of what the key is for, as it will be the only way of identifying where this key is used later. The role dropdown is used to set the permissions of the API key, and like users a key can be a viewer, an editor, or an admin. It's best practice to use the lowest level of permissions required, so if you're using this key to retrieve information from Grafana but not to add or change anything, stick with the viewer role; you can always add another key later for editing. Finally, you can optionally specify a duration for the key, such as "30d" for 30 days or "8h" for 8 hours. After this time period, the key will expire and no longer function. Leaving this field blank will create a key that is valid forever.

Figure 13-2. *Creating an API key*

After you configure the values for your API key, click the "Add" button to finish creating it. You'll then see the API key that was created, shown in Figure 13-3. The value in the "Key" field is the actual API key. Be sure to copy the entire key, as it may be too long to fit entirely in your browser window and may have characters at the beginning or end that aren't automatically selected by a double-click.

Figure 13-3. *A newly created API key ready to be used*

You'll also see an example command line using the key at the bottom of this view. This can be used with the `curl` utility to verify that the key is configured and working correctly when used with your Grafana environment. Testing your key this way is completely optional; skipping this will not prevent your key from working.

Caution   This is the **only** time you'll see your API key in Grafana, so be sure to copy it right away! There's no way to retrieve it again after this; if you don't copy the key or miss part of it, the only option is to delete the key and create a new one.

Deleting API Keys

To delete an API key, click the red X on the right next to the key information. You'll be asked to confirm that you really want to delete the key before it will be deleted. Once you confirm this, the key will be immediately removed from Grafana and no longer valid.

Using API Keys

Grafana API keys are OAuth 2.0 bearer tokens. This means that in order to work, they need to be included as a header in every HTTP request. This is shown in the example command that is generated when you create a key, like the one at the bottom of Figure 13-3. Each request made to Grafana using an API key will need a section in the header of the request like

```
Authorization: Bearer <token>
```

If you're using a REST API client, you'll need to be sure that this header is included in your requests. Without it, you'll get a 403 Forbidden error from Grafana when trying to access the API.

To add this header in Postman, create a new collection for Grafana. In the collection authorization tab, select the bearer token type and add your Grafana API key as the token. Figure 13-4 shows the configuration of a Grafana API key in a Postman collection.

Figure 13-4. *A Postman collection with the required API key fields highlighted*

After setting this header for the collection, any requests that you create inside the collection will automatically include this header. (Be sure to click the save button so that the settings are updated. If you try to create requests before saving, the header will not be applied.)

To test that your key is working, you can use the URL from the `curl` command provided when you create your API key, which is *https://<your grafana instance>/api/dashboards/home*. Create a GET request to this URL and run it, and you should see a result similar to the one in Figure 13-5.

Figure 13-5. *A successful request to the dashboards API returns dashboard information*

If instead you get a 403 error or see a message like

```
{
    "message": "Unauthorized"
}
```

then either there is something wrong with your API key (such as a copy/paste error) or the header is not being properly applied to your request. In this case, you'll need to do some troubleshooting with your REST client until you resolve the issue.

The Grafana API

In this section, we'll take a brief look at the Grafana API. This API controls almost every aspect of Grafana from configuration to operation and even monitoring the health of Grafana itself. To cover the whole API would require a book on its own, so we'll touch on the basics and some of the more useful functionality here. The full API is documented on the Grafana website at *https://grafana.com/docs/grafana/latest/*.

Before diving into how to use the API, it's important to know a bit about how Grafana represents resources internally. Almost everything in Grafana – dashboards, folders, data source configuration, you name it – is represented in JSON format. JSON is a *data serialization* format, which is a fancy way of saying that it can represent complex, multidimensional data structures in a simple form.

When you create a dashboard in the Grafana UI, Grafana is representing that dashboard as a JSON object in the background. So when you use the API to retrieve or update a dashboard, you'll be seeing that JSON format a lot.

The good news is that JSON is just a text format, so working with it is like working with any other plain text. You can save it, edit it, or check it into source control as you like, and there are a number of tools available for working with it directly. Having JSON as a format for all Grafana data makes it easy to connect to other systems or back up data wherever you prefer.

Dashboards

The dashboards API allows you to manage Grafana dashboards. Using this API, you can get, update, create, or delete dashboards and retrieve some metadata about dashboards such as the tags that have been applied to it.

Getting Dashboards

To retrieve the JSON object that defines a dashboard in Grafana, you'll need to know the dashboard's unique identifier (UID). This is the string of characters after /d/ in your dashboard URL up to the following slash. For example, if the URL of a dashboard in Grafana is

```
https://gettingstartedwithgrafana.grafana.net/d/glLNuda7z/my-
dashboard?orgId=1
```

then the dashboard's UID is glLNuda7z.

319

To get the dashboard's JSON, you can put this UID into the dashboards API and use an HTTP GET on the URL:

```
https://<your grafana instance>/api/dashboards/uid/<dashboard UID>
```

This will retrieve the JSON representation of the dashboard with the given UID. An example of this is shown in Figure 13-6.

Figure 13-6. *Getting a dashboard by using its UID returns the JSON representation of the dashboard content*

Using this API, you can download a dashboard's configuration and check it into source control or otherwise back it up and manage it as code. This can then always be imported using the create or update API (described below) to add the dashboard back into a Grafana environment.

Creating or Updating Dashboards

Creating and updating a dashboard in Grafana use the same API endpoint. If a dashboard with the specified UID exists, it will be replaced with the one you send to the API. If it doesn't already exist, it will be created.

To create or update a Grafana dashboard, you will POST data to the API at

```
https://<your grafana instance>/api/dashboards/db
```

Note that the URL does not contain the UID of the dashboard; this is set in the JSON object that is the body of the request. You'll need to ensure that the content-type header of your request is set to application/json in order for Grafana to accept your dashboard. This should be done in your REST client if you're using one. In Postman, this is set in the details of the "body" tab of the POST request. Figure 13-7 shows the result of a successful POST of a dashboard to the API.

Figure 13-7. *Using HTTP POST to send a dashboard definition (top text box) to Grafana, and the resulting success message (bottom text box)*

If you attempt to use an API key with the *viewer* role, you'll receive a result like

```
{
    "message": "Access denied to save dashboard"
}
```

If you see this, check your API key. To update or create a dashboard, you'll need to use a key with the *editor* or *admin* role.

Deleting Dashboards

Deleting a dashboard uses the same URL structure as getting a dashboard:

```
https://<your grafana instance>/api/dashboards/uid/<dashboard UID>
```

The difference is the HTTP verb used. Instead of GET, use a DELETE.

Caution   As soon as you execute the DELETE API call, the dashboard will be deleted. There's no confirmation here, so be sure that you have the correct ID!

Successfully calling a DELETE on the API will give you a success message like

```
{
    "id": 18,
    "message": "Dashboard MyDashboard deleted",
    "title": "MyDashboard"
}
```

If you attempt to use an API key without permission to delete dashboards, such as a key with the *viewer* role, you'll instead get a failure message like

```
{
    "message": "Access denied to this dashboard"
}
```

In this case, check that you are using a key with *editor* or *admin* permissions.

Folders

The folders API lets you manage dashboard folders in your Grafana environment. This includes listing folders, getting and setting the title of folders, and creating and deleting folders.

Listing Folders

To list folders, use an HTTP GET on the folders API, like

```
https://<your grafana instance>/api/folders
```

You can optionally add a limit on the number of folders that are returned by this API call by adding the limit parameter. For example, adding ?limit=10 to the end of the URL will return only the first ten dashboards. If you don't specify a limit, the default limit of 1000 will be used.

This is most useful for finding the unique identifier (UID) of a folder that you want to delete or update. Like with dashboards, the UID is used to tell Grafana which specific folder you want to change or delete.

Figure 13-8 shows an example of listing the folders in a Grafana instance using Postman. The full list of folders for this environment is shown in the JSON output in the bottom of the screen.

Figure 13-8. *Getting all dashboards in a Grafana environment returns dashboard UIDs and names*

Getting Folder Information

To get the metadata attached to a single folder, you can pass the folder UID into the API call using an HTTP GET:

```
https://<your grafana instance>/api/folders/<folder UID>
```

This will return much more information about a dashboard than the name and UID that the dashboard listing API gives. When getting the information about a specific folder, you'll also see details about when and by whom the folder was created and last updated, permission information, and the full URL to access the folder in Grafana. Figure 13-9 shows an example of calling the folder API to get information about a specific folder.

Figure 13-9. *Viewing a specific folder gives more detailed information than is available in the list of all folders*

Creating Folders

To create a folder, use an HTTP POST to the folder API:

```
https://<your grafana instance>/api/folders
```

The body of the request should contain a JSON description of the folder. The only required field in this JSON is the folder title; you can optionally add a UID if you want to specify the UID yourself. If you don't include the UID, one will be automatically generated and returned in the response.

Figure 13-10 shows an example of creating a folder using Postman. Be sure that the content-type header of your request is set to application/json so that Grafana knows to expect JSON data; in Postman, this is done by picking the content type when setting the POST options. The results of the API call provide information about the folder that has been created.

Figure 13-10. *Using HTTP POST to create a folder. The folder title (top text box) is sent to Grafana and the resulting created folder (bottom text box) is returned*

If you try to create a folder with the same name as an existing folder or dashboard, you'll get this response:

```
{
    "message": "a folder or dashboard in the general folder with the same
    name already exists"
}
```

In this case, you'll need to delete or rename the existing folder or dashboard or pick a new name for the folder you're creating.

On the other hand, if you try to create a new folder with the same UID as an existing folder, you'll get this message:

```
{
    "message": "the folder has been changed by someone else",
    "status": "version-mismatch"
}
```

Unlike dashboards where create and update is the same API call, folders have a separate API for each of these actions. To update a folder, see the following section on updating folders.

Finally, if you try to add a folder but don't have permission to write data to Grafana, you'll get a message like

```
{
    "message": "Access denied"
}
```

In this case, make sure that you are using an API key with *editor* or *admin* privileges and try again.

Updating Folders

While dashboards use the same API call to create and update data, folders have a separate API call for each. To update a folder, you'll need to put its UID into the folder API and make an HTTP PUT:

```
https://<your grafana instance>/api/folders/<folder UID>
```

In the body of your request, include the data that you want to update. You can change the UID or the title of the folder; the other metadata is managed by Grafana directly.

When updating a folder, you'll need to either increment the *version* of the folder or pass a special *overwrite* parameter in as part of the body of the request. Without doing one of these, you'll get a response like

```
{
    "message": "the folder has been changed by someone else",
    "status": "version-mismatch"
}
```

Grafana will not let you update a folder without either setting a new version or explicitly setting overwrite to true. Generally, the latter is easiest, as otherwise you'll need to request the existing version number and update it yourself when sending data to Grafana. Figure 13-11 shows an example of updating a folder title using the overwrite property. Note that like creating a folder, you need to ensure that the content-type header of your request is set to application/json. In Postman, this is set when creating the request body.

Figure 13-11. *When updating a folder, the request (top text box) needs to contain a version number or the overwrite property. Results (bottom text box) include the new metadata for the folder*

Deleting Folders

Deleting a folder in Grafana involves sending an HTTP DELETE to the folder API and passing in the folder UID:

```
https://<your grafana instance>/api/folders/<folder UID>
```

Successfully calling a DELETE on the API will give you a success message like

```
{
    "id": 20,
    "message": "Folder my new folder deleted",
    "title": "my new folder"
}
```

Deleting the folder requires *editor* or *admin* permissions for your API key. If you try to delete a folder with viewer permissions, you will receive a message like

```
{
    "message": "Access denied"
}
```

If you see this, change to an API key with the appropriate permissions and try again.

Caution   Deleting a folder happens immediately when you run an HTTP DELETE against the folder API. This will delete the folder and all dashboards inside it and **cannot be undone**. Be sure you have the right UID before calling the API!

Data Sources

The data sources API lets you manage the configuration of data sources within Grafana. This API is a bit more flexible than the API endpoints we've seen earlier. In addition to working with unique identifiers (UIDs), they can also use data source names or Grafana IDs. For consistency, we'll continue to use UIDs in the examples here (with one exception, updating; see the updating section for details). If you want to see the other ways to manage data sources via an API, check out the Grafana API documentation.

All uses of the data sources API require an API key with *admin* permissions. If you try to use a key with *viewer* or *editor* permissions, you'll get an error:

```
{
    "message": "Permission denied"
}
```

If you see this, switch to an admin API key and try your request again.

Listing Data Sources

To list data sources, use an HTTP GET on the data sources API endpoint:

```
https://<your grafana instance>/api/datasources
```

This will return a list of all data sources configured in your Grafana instance as shown in Figure 13-12.

Figure 13-12. *Listing data sources via the Grafana API*

Tip   For security reasons, the output from the data sources API does not show credential information like passwords. This means that it's generally safe to store the results in source control, but you should still check the output before adding anything potentially sensitive to source control, an email, etc. If credentials are embedded in a URL or other text field, they could still be included in the output here.

Getting Data Source Information

To retrieve information about a specific data source only, you can use an HTTP GET on the data sources API and pass a specific UID:

```
https://<your grafana instance>/api/datasources/uid/<data source UID>
```

This will return the same information as the data source list, but will limit the results to only the data source with the UID provided.

Creating Data Sources

To create a data source, use an HTTP POST to the dashboards API:

`https://<your grafana instance>/api/datasources/`

In the body of the request, include a JSON representation of the data source configuration. (The easiest way to get this is to export a data source using the GET data sources API (mentioned earlier) and modify it to contain the values that you need.) You don't need to include a UID in this – if you leave it out, Grafana will create one for you and return it in the results.

You'll also need to set the content-type header of your request to application/json so that Grafana knows to expect JSON data; in Postman, this is done when setting the POST options. Figure 13-13 shows the result of creating a new data source via the API.

Figure 13-13. *Creating a data source based on a definition (upper text box) passed to the Grafana API. The resulting data source, including the new UID, is provided in the result (lower text box)*

Tip If your data source requires credentials, such as a username and password, be sure to include them in the request! They will be encrypted before being stored in Grafana and won't be shown in the response, so your data will be kept secure.

If you try to create a data source using a UID or a name that already exists, you'll get a response like

```
{
    "message": "data source with the same name already exists"
}
```

In this case, you'll need to either delete the existing data source or change the UID or name of the data source that you're creating. (If you want to update a data source, see the following section on updating.)

Updating Data Sources

To update a dashboard, use an HTTP PUT to send the updated dashboard to the API:

```
https://<your grafana instance>/api/datasources/<data source ID>
```

Unlike most of the other parts of the data sources API, the update API requires you to use the Grafana ID number rather than the UID or name of the data source. You can find this by using the GET data sources API outlined earlier.

The body of the request follows the same rules as for creating a data source; the only difference is that this API call will not create a data source if a data source with the provided ID doesn't already exist.

Deleting Data Sources

To delete a data source, send an HTTP DELETE to the data sources API:

```
https://<your grafana instance>/api/datasources/uid/<data source UID>
```

This will delete the data source immediately and return a success message:

```
{
    "message": "Data source deleted"
}
```

> **Caution**   Like other delete APIs, there are no prompts for confirmation here. If you delete a data source that is in use by a dashboard, that dashboard will not show any data that would normally come from the deleted data source. Be certain that you are deleting the correct data source when using this API.

Miscellaneous

There are a large number of API calls that haven't been covered here, so if there are things in Grafana that you want to automate, it's worth looking through the full API documentation. That said, there are a couple of other generally useful API calls that don't fall into the preceding categories that are worth mentioning.

Grafana Health

There's an API endpoint that exists to show the current status of your Grafana environment. This can be useful to monitor the state of Grafana itself and alert you if there are issues in your monitoring environment.

To check the state of Grafana, make an HTTP GET to the health API:

```
https://<your grafana instance>/api/health
```

This will return some brief status information about your Grafana instance:

```
{
    "commit": "53f5c6a44",
    "database": "ok",
    "version": "8.4.1"
}
```

URL Shortener

Grafana contains a URL shortener. This can be used to make short, easy to copy and paste links to dashboards and other resources. (The URL shortener will only work for relative paths in the Grafana environment, so it's Grafana specific.) This is particularly useful when you have a dashboard link with many variables or custom time ranges set, which can lead to a very long URL. Using the URL shortener, you can make a version of this URL that leads to the same dashboard in Grafana but doesn't take too much space to paste in an email.

To use it, send an HTTP POST to the URL shortener API:

```
https://<your grafana instance>/api/short-urls
```

In the body, include the path to shorten as a JSON object, being sure to the content-type header of your request to application/json so that Grafana knows to expect JSON data; in Postman, this is done when setting the POST options:

```
{
  "path": "<your URL>"
}
```

Figure 13-14 shows a request shortening a long URL to a shorter one.

Figure 13-14. *A longer URL provided in the request body (top text box) and the resulting short URL as the request response (bottom text box)*

The URL does have to be a relative URL, meaning that the server part of the initial URL must be the Grafana server you're calling this API on. If you try to pass a full URL or one starting with a /, you will get a message like

```
{
    "message": "Path should be relative"
}
```

If you see this, shorten your URL to a relative one and try again.

Grafana Provisioning

The Grafana API is a great way to automate your Grafana environment once it is up and running. But if you want to completely automate the configuration of Grafana from initial installation through full production, the provisioning system is the way to go.

Like with the API, this section won't be a full exploration of the full functionality of provisioning. That is available in the Grafana documentation at *https://grafana.com/ docs/grafana/latest/administration/provisioning/*. This section will introduce you to how the provisioning system works and provide some pointers for managing Grafana through it.

Provisioning Overview

The Grafana provisioning system is a set of configuration files and locations on disk that tell Grafana how to configure itself. Through it, you can define where Grafana should look for configuration files that describe the layout of dashboards and folders and set up data sources.

The provisioning system is controlled in *grafana.ini*, the main Grafana configuration file. It's enabled by default, meaning that you can put configuration files in the appropriate location with no additional changes to Grafana, and the system will be provisioned when you start it up. By default, configuration files are in *<your grafana path>/conf/provisioning*, but this can be changed in grafana.ini.

Resources configured via the provisioning system behave differently inside of Grafana. Normally, anyone with the appropriate role can change or delete a dashboard or data source in Grafana, regardless of how it was created. But provisioned resources can only be changed through the provisioning system – attempting to change or delete them in the UI or via the API will fail and give a message telling you that you need to reprovision the resource to make a change. This means that provisioning is also a great way to make special protected dashboards and data sources in the case where you want to prevent something from being changed manually.

Finally, because provisioning is managed by putting configuration files in specific locations on disk, this means that it is not accessible via Grafana Cloud. Provisioning is only useful in a self-managed Grafana environment.

Data Sources

To provision data sources, you will add a configuration file for each data source to the data source provisioning path. By default, this path is *<your grafana path>/conf/ provisioning/datasources* but can be changed in the grafana.ini file.

Configuration files for data sources are in YAML format. YAML is a data serialization format, similar to JSON, but is easier for humans to read.

Each data source that you want to provision will have its own configuration file. The file needs to contain all of the information that you would normally add in the UI when creating a data source, so things like the data source name, type, location, and credentials. Each field is a text string with an identifier followed by the value. A sample file with the commonly used fields is in the data source provisioning path by default and is a good starting point.

Let's look at an example data source provisioning file in Listing 13-1. This file contains everything needed to set up a MySQL instance in Grafana.

Listing 13-1. A data source provisioning configuration for MySQL

```
apiVersion: 1

datasources:
  - name: MySQL
    type: mysql
    url: mysql.example.com:3306
    database: web_analytics
    user: grafana
    secureJsonData:
      password: grafana
```

The first line, setting the API version, is used to tell the provisioning system what version of the provisioning format it is using. Currently, there is only one version, so this will always have a value of 1.

The next section, *datasources*, defines a list of data sources that are configured here. Each new entry in the list is indented and starts with a dash (-). In this case, our list contains only one data source, but it's possible to configure more than one in a single file.

Inside of this list are fields representing the data that would normally be typed in the UI when creating a data source. We tell Grafana what to call the data source, what type it is, where it's located, etc.

You'll notice that the username and password are kept in this file. The password itself is marked as something that should be encrypted when loaded into Grafana by being part of the *secureJsonData* section, but it is still visible to anyone with read access to this provisioning file, so bear that in mind – you'll want to limit access to the internals of your Grafana environment if you're using provisioning.

Once this file is saved in the appropriate location, Grafana will configure this data source when starting up.

Dashboards

Similar to data sources, you can tell Grafana to import dashboards from configuration files on disk. But because dashboards are in JSON rather than YAML (and can be potentially quite long!), the full dashboard data isn't copied into the configuration file for provisioning.

Instead, the provisioning file for dashboards will tell Grafana where to look for the actual dashboard JSON files themselves. You provide a path on disk, and Grafana will load each file from that path as a dashboard. You can also set a time period for Grafana to look for updates, and it will check those files for changes on the schedule you set. This lets you update provisioned dashboards without having to restart your Grafana environment for the changes to be picked up.

Listing 13-2 is an example of a dashboard configuration file. Just like with data source provisioning, dashboard provisioning uses YAML as its format.

Listing 13-2. Provisioning dashboards in Grafana lets you point to folders full of dashboard definitions easily

```
apiVersion: 1

providers:
- name: 'Web monitoring'
```

```
orgId: 1
folder: 'Web monitoring'
type: file
updateIntervalSeconds: 60
options:
  path: /opt/grafana/dashboards/web
```

This file should look pretty familiar. Like with data sources, we start by telling Grafana what version of the API we're using, which is still 1.

The *providers* section contains a list of places on disk for Grafana to find dashboards. Grafana will look in the location specified in *path* and load all JSON files it finds there as dashboards. These should be normal dashboards that have been exported via the Grafana UI or retrieved through the API. This path does not have to be inside of your Grafana deployment location. Any accessible path on your server will work.

The *folder* option tells Grafana what dashboard folder these dashboards should be part of. If the folder doesn't exist, it will be created for you.

Finally, *updateIntervalSeconds* tells Grafana how frequently to look for changes in these files. In this case, we're looking once per minute for updates.

Summary

In this chapter, we covered the Grafana API and provisioning system. You learned how to create API keys and assign them roles. You saw how to use the API to manage resources in Grafana like dashboards, folders, and data sources. We also looked at the Grafana provisioning system which allows you to configure these resources via files on disk.

In the next chapter, we'll take a look at the additional features available in Grafana Enterprise.

CHAPTER 14

Grafana Enterprise

Up to now, everything we've looked at has been part of the standard, open source version of Grafana and available for free. In this chapter, we'll explore features that are part of Grafana Enterprise, the commercial version of Grafana. In order to use these features, you'll need a valid Grafana Enterprise license, either for your Grafana Cloud environment or to run in your own environment.

A major component of Grafana Enterprise is support and indemnification. If visualization and alerting on the metrics in your environment is critical to your business, it makes sense to have 24/7 support and protection in the event of any legal issues that arise in your environment. For some organizations, this alone may be worth the price.

Beyond support and legal coverage, Grafana Enterprise extends standard Grafana in a few key ways. It provides additional connectors to data sources that aren't part of the open source Grafana platform. It provides additional security features that let you limit access to data and data sources, set more fine-grained access controls to features of Grafana, and map those roles as well as team memberships from external authentication systems – and ensure that those changes stay live and in sync across Grafana instances. Scheduled PDF reporting, dashboard usage and viewership information, and enhanced searching of dashboards by most/least useful help you collaborate with colleagues more easily. Finally, Grafana Enterprise gives you the option to change the appearance of Grafana itself to more closely match your organization's style.

Because Grafana Enterprise is a paid product, we'll only look at some of the highlights here – to fully understand and use Grafana Enterprise, you will need to sign up and pay for a subscription. Once you do this, you can contact Grafana Labs for any help that you need, as support is part of this subscription.

© Ronald McCollam 2022
R. McCollam, *Getting Started with Grafana*, https://doi.org/10.1007/978-1-4842-8309-7_14

Enterprise Data Sources

The first – and probably simplest – feature of Grafana Enterprise is the addition of data sources that are not a part of the standard open source Grafana environment. As a general principle, if a potential data source for Grafana is itself open source, the data source plugin in Grafana likely will be as well. But there are a number of proprietary data stores that could be useful to connect to Grafana, and when a plugin for one of these exists, it's most likely to be part of Grafana Enterprise.

For example, in Chapter 4 we looked at connecting popular open source relational databases MySQL and PostgreSQL to Grafana. But while Oracle is also widely used, it's not open source themselves. So in this case, the plugins for the Oracle database are part of the Grafana Enterprise suite.

Note   This isn't a hard and fast rule. Some data source plugins, such as Microsoft SQL Server, are available in the open source edition of Grafana. You can use this as a rough guideline when thinking about data sources, but always check the official list on *https://grafana.com/grafana/plugins/?type=datasource* to be sure.

Aside from requiring a license, Grafana Enterprise data sources do not behave any differently than other data sources in Grafana. They may have their own specialized query editors to make querying the data easier, as shown in Figure 14-1 for the Dynatrace data source, but they always function just like any other data source.

Figure 14-1. *Enterprise data sources, such as Dynatrace, may have their own query builders but behave like any other data source in Grafana*

Enterprise data sources can be used alongside open source data sources in Grafana – they behave just like any other data source and have the same panel types, options, and configuration as other data sources.

And just like open source data sources, Grafana Enterprise data sources are provided as plugins for Grafana. At the time of writing this book, the list of available Enterprise data sources includes the following:

- **Databases:** MongoDB, Oracle, SAP HANA, Snowflake

- **Logs and metrics:** AppDynamics, Datadog, Dynatrace, Honeycomb, New Relic, Splunk, SignalFx, Wavefront

- **Other tools:** GitLab, Jira, Salesforce, ServiceNow

A full up-to-date list of Enterprise data sources can always be found on the Grafana website at *https://grafana.com/grafana/plugins/?enterprise=1*.

Enterprise Security

While the open source edition of Grafana provides the ability to restrict read and write access to specific dashboards and folders, this may not be enough coverage for organizations with more advanced security requirements. Grafana Enterprise provides some additional features to enhance the Grafana security model.

Data Source Permissions

Limiting read access to dashboards is a good start for restricting access to sensitive data, but it doesn't solve every possible problem. For example, anyone with access to the Grafana explore view can select any data source in Grafana and run queries against it.

But even removing access to explore doesn't completely address the issue. Consider the case of a dashboard that shows sensitive information that should not be generally available. It can be placed in a dashboard folder with restricted permissions that allow only those users who have been explicitly added to that folder to see the sensitive dashboard.

Now let's imagine that one of the users of this dashboard decides to make some necessary improvements to the layout. Because they're experimenting with a dashboard that is in production, they decide to first make a copy of that dashboard to a folder that they've created and work on the copy. Unfortunately, they didn't properly restrict access to their working folder, and now anyone with a Grafana account can potentially view the sensitive data.

So even without malicious intent, it's possible for sensitive data to leak in a Grafana environment. To combat this, we can use *data source permissions* in Grafana Enterprise.

In the same way that dashboard and folder permissions control who has access to those resources, data source permissions restrict who can query a specific data source. Once enabled on a data source, anyone who does not have query permissions explicitly enabled for their account or a group their account is a member of will not see that data source at all.

When Grafana Enterprise is licensed, a new tab will appear at the top of each data source configuration page which lets you control data source permissions. By default, these permissions are disabled (meaning anyone can see each data source, exactly as in open source Grafana). Figure 14-2 shows the default state of data source permissions. Clicking the *enable* button will turn permissions on for this data source.

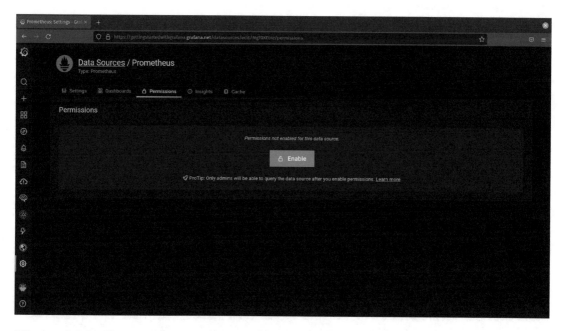

Figure 14-2. *Data source permissions are disabled by default. Enabling them will restrict access to this data source*

After enabling permissions for a data source, you can explicitly add users or teams who have access to query that data source. Figure 14-3 shows data source permissions enabled on the Prometheus data source, allowing only a specific user account to access the data source. (Note that administrators can always access any data source; this permission cannot be changed.)

Figure 14-3. *The data source permissions panel allows an administrator to explicitly grant access to query a data source*

Once data source permissions are enabled, any user that does not have access to the data source will not be aware of its existence. The data source will not show up as an option for querying in panels or through the explore view. Furthermore, any dashboards that query this data source will be missing the data source entirely.

For example, in Figure 14-4, a dashboard has been configured to query temperature data from two sources, Prometheus and InfluxDB. In the browser on the left, a user with permission to access both data sources is logged in. In the browser on the right, a user without access to the Prometheus data source is logged in. The user without Prometheus access sees only a blank panel here.

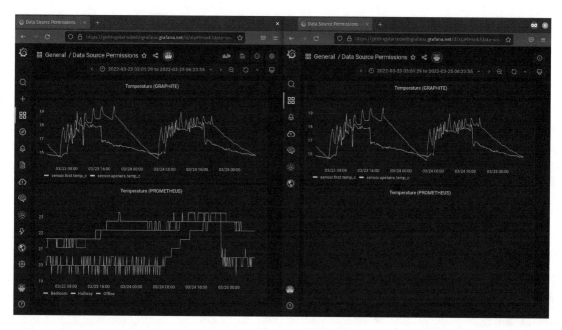

Figure 14-4. *Users with access to a data source will see its data in a dashboard (left), while users without access will have a more restricted view (right)*

Fine-Grained Access Control

Open source Grafana provides three roles for users: view, edit, and admin. In more complex environments with deeper security requirements, these roles may not provide enough granularity to properly capture user permissions.

Grafana Enterprise extends this capability by adding the notion of fine-grained access control. In this model, you can create your own roles and assign specific access permissions to that role. For example, you might want to restrict access to the explore view to only authorized users to help head off data leaks as described earlier under data source permissions. You could turn off access to the explore view by default and add access to it to a role that can be assigned to a user or a team.

The flexibility of this sort of system also means that it has a large number of possible options. So for more information about roles and permissions in Grafana Enterprise, consult the documentation at *https://grafana.com/docs/grafana/latest/enterprise/access-control/permissions/*.

Enterprise Access and Authorization

In Chapter 12, we looked at ways of controlling access to Grafana from external systems such as OAuth and LDAP. Grafana Enterprise offers more functionality and extends some of that access and authorization control.

Enhanced LDAP

Open source Grafana can be connected to an LDAP directory for the purposes of enabling logins from a central directory service. However, this functionality has some limitations: while roles can be mapped from LDAP group membership, team membership inside of Grafana cannot. And permissions are checked only when someone logs in, meaning that if their access is changed (or revoked!), then nothing will take effect until they log out and log back in. Grafana Enterprise addresses these limitations.

LDAP Team Sync

LDAP team sync enables you to automatically map members of a group in your LDAP environment to a team in Grafana. This means that anyone who is a member of a group that has been mapped in this way will automatically be made a member of the appropriate team inside of Grafana, even if this is their first time logging in – no additional configuration or manual mapping of the user to the team is needed.

Figure 14-5 shows an example of mapping a group in LDAP to a team in Grafana, accessed by opening the team configuration page in Grafana. Note that you need to specify the full name of the group to map in your LDAP environment, which is not necessarily the same as the friendly name that is displayed in many user interfaces. In this case, the Scientists group in LDAP is actually named ou=scientists,dc=example,dc=com in the directory service. Be sure to check that you have the full group name when creating these mappings.

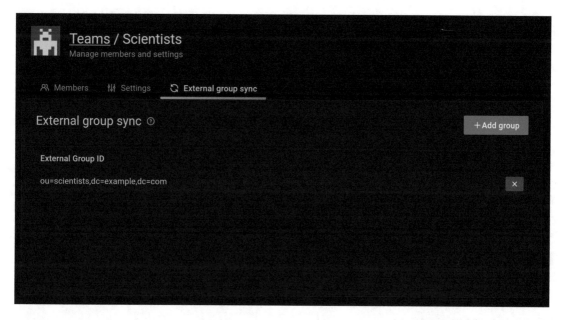

Figure 14-5. *Mapping an LDAP group to a Grafana team requires the full group name from the LDAP directory*

LDAP Active Sync

LDAP active sync tells Grafana to update LDAP access and group membership on a regular, scheduled basis. This means that changes will be applied as soon as the scheduled update completes, even for logged in users.

To enable LDAP active sync, you will need to edit your Grafana configuration file, *grafana.ini*. Find the LDAP authorization section headed by [auth.ldap]. Inside this section, you will configure the schedule that Grafana will use to synchronize to your LDAP server. The schedule is set using Unix cron formatting, which specifies fields for second of the minute, minute of the hour, hour of the day, day of the month, month, and day of the week, any of which can contain a wildcard character *. Listing 14-1 has several examples of schedules that can be used.

Listing 14-1. Configuring LDAP active sync schedules in grafana.ini

```
[auth.ldap]
# Format is:
# [second] [minute] [hour] [day of month] [month] [day of week]
```

```
sync_cron = "0 0 1 *  * *" # default, daily at 0100

# Other possible options:
# sync_cron = "0 0 * * * *"    # hourly at X:00
# sync_cron = "0 30 3 * * 0"   # weekly at 0330 on Sunday
# sync_cron = "0 */10 * * * *" # every 10 minutes
```

Once you make these changes, save your configuration file and restart Grafana for them to take effect.

SAML Support

Similar to OAuth and LDAP authentication, SAML is a type of single sign-on (SSO) authentication. Grafana Enterprise provides the ability to authenticate users, connect to resources, remotely manage user permissions (and even log them out remotely), and many other functions.

Configuring SAML in Grafana Enterprise is similar to configuring OAuth or LDAP, involving editing a section of the Grafana configuration file, *grafana.ini.* However, SAML support in Grafana Enterprise provides far more options and exposes a much greater granularity of control when compared to these other authentication mechanisms. For a full description of SAML support in Grafana Enterprise, consult *https://grafana.com/ docs/grafana/latest/enterprise/saml/.*

Scheduled Reporting

Grafana excels at showing real-time data and letting viewers interact with it. But sometimes it can be useful to take a static snapshot of data at regular intervals. You might want to keep historical data in an offline, read-only repository for regulatory purposes. Or maybe the CEO wants to stay updated on what's happening across the organization but doesn't want to have to remember to log in to Grafana to check manually.

These sorts of use cases are what the Grafana Enterprise scheduled reporting functionality is built to address. You can set up a report to run at regular intervals that will be emailed to one or more recipients as a PDF attachment. Think of this like taking a screenshot of a dashboard and emailing it without the hassle of doing it by hand.

Once you have a valid Grafana Enterprise license, you'll see a new menu item on the navigation bar on the left which will take you to the report management view. Figure 14-6 shows the report management view as well as the navigation menu that is used to access it. From here, you can create, edit, or delete reports.

Figure 14-6. *Grafana Enterprise adds a scheduled reporting feature which is managed in the Grafana user interface*

Reports expose a number of options to let you get things just the way you want them. Here, you can pick a dashboard that you want to use as the base for the report and, as shown in Figure 14-7, configure how this should behave. This includes the recipients that should get a copy of the report, a message to attach to the email, and a number of options for how the report should look and behave.

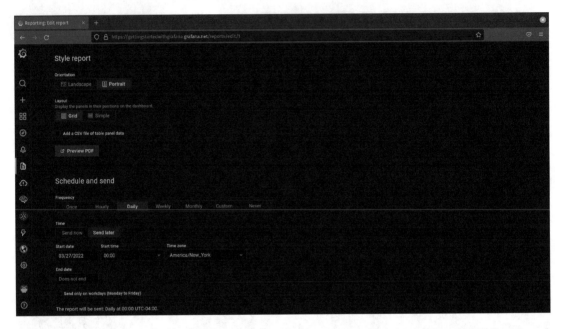

Figure 14-7. *Scheduled reporting exposes a number of options to control how and when a report is generated*

Most of the options are self-explanatory, but a few are worth calling out. The orientation option determines if the PDF will be in portrait mode with a page longer than it is wide (standard for printed pages) or landscape mode with a page wider than it is tall (more normal for dashboards on televisions or other large screens). Using *grid* mode will lay the dashboard out in a grid exactly as it would be in a browser, but changing this to *simple* will set up all the panels on the dashboard one per row from the top of the page to the bottom. (This is useful in cases where you may have more data than can comfortably fit in a small page when rendered to PDF.)

Finally, you have a number of options to configure how often the report will be generated.

Tip   If you've configured an external image rendering service as described in Chapter 12, this will be used when creating reports. Otherwise, your Grafana instance will generate all the reports itself. If you have a large number of reports and/or users, consider using an external image renderer so that your Grafana instance does not see a performance impact.

Dashboard Insights

Another new icon you might notice when you upgrade to Grafana Enterprise is the dashboard insights button at the top of each dashboard, highlighted in Figure 14-8.

Figure 14-8. *Grafana Enterprise adds a button to each dashboard to access dashboard insights*

Dashboard insights provide a view into how dashboards are being used and how they're performing in your environment. Clicking this button on a dashboard will bring up a view of insights data for that dashboard specifically. This includes things like frequency of access, number of errors that have been seen on that dashboard, and information about who is editing and viewing it over time. Figure 14-9 shows some of this information on a single dashboard.

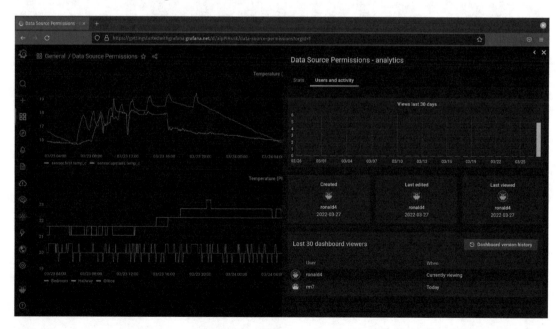

Figure 14-9. *Dashboard insights for a single dashboard can show interesting usage information*

While usage insights on a single dashboard can be interesting, their real power is shown when looking at aggregate data. This can help you understand which dashboards in your environment are most (and least) popular as well as which dashboards most frequently encounter errors. Using this data can help you improve your users' experience and provide even more value to consumers of your data.

To access the aggregate dashboard insights data, start by browsing all dashboards. You can then filter by aggregate dashboard insights data as shown in Figure 14-10. (Remember that you can always click one of these dashboards and view the individual dashboard insight data to get more details!)

Figure 14-10. *Aggregate dashboard insights data can help you understand the best – and worst – aspects of your users' experience in Grafana*

Changing Grafana's Appearance

Grafana Enterprise offers one more major change over the open source Grafana: the ability to change much of the appearance of Grafana itself.

For many organizations, having a consistent look and feel is a major part of who they are. This is not only for consistency in branding but can be an important marker of legitimacy. An application that is integrated into a platform feels like it belongs, where a transition to another color scheme, set of icons, and naming can be jarring.

Grafana Enterprise allows Grafana admins to customize much of its appearance. This includes things like the title of the application that is displayed in the browser, adding additional links to services or information in the Grafana footer menu, and even changing the login screen and the system icon used at the top of the left navigation bar.

The specifics of what can be changed and how it is managed varies depending on the version of Grafana used and is evolving over time. A full list of the options that can be changed is available at *https://grafana.com/docs/grafana/latest/enterprise/ white-labeling/*, and Figure 14-11 shows some of the changes that are possible.

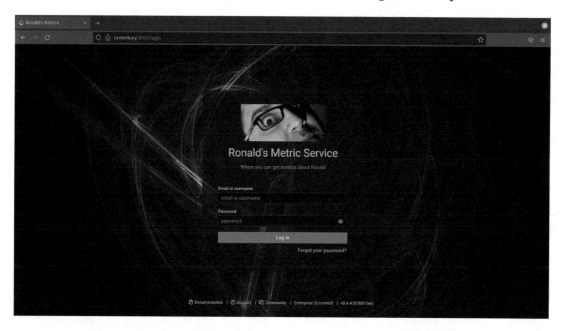

Figure 14-11. *Grafana can be rethemed to match your organization's branding, no matter how strange that branding is*

Summary

In this chapter, we've taken a quick tour of the functionality available in the paid version of Grafana, Grafana Enterprise. You've learned about additional data sources that can be added to connect Grafana to other repositories of data in your environment. We reviewed more of the advanced authentication and access methods that can be applied, including additional ways of connecting to Grafana and ways to add tighter control over access to the resources inside of Grafana. You have seen how to limit access to data sources based on user account or team membership, and we touched briefly on extending this sort of granular access control through Grafana Enterprise's fine-grained access control.

You've also learned about dashboard insights, a way of tracking the use, editing, and experience that users have in your Grafana environment. Finally, we took a quick tour of some of the changes that can be made to Grafana's appearance to make it match your organization's style more closely.

Index

A

Account creation
  administrator, 6
  email address, 4, 5
  home page, 4
  login page, 4
  navigation bar, 6
  restrictions, 5
  SSO system, 5
  supported service, 4
  URL, 5
  validation, 5
  web page, 4
"Actual" button, 22
Add data source, 91
Adding/removing users, 133–139
Additional context, 157, 165–167, 194, 281
"Add link" button, 182, 186
"Add permission" button, 146
Address validation, 5
Ad hoc queries, 19
Administrators, 6, 49, 51, 56, 60, 73, 79,
    180, 300
"Admin" user account, 149
Alerting system, 265
  conditions, 279, 280
  configuration, 266
  contact points, 268–270
  create, 275–277
  details, 280, 281
  expression, 267
  message templates, 268

query, 278
rules, 266, 267
rule type, 277
state, 267
Alert rules, 267
Alert tab, 23
Amazon, 4
Amazon RDS, 114
API keys, 107, 122, 313
  add, 314
  create, 315
  delete, 316
  using, 317, 318
API Tokens control panel, 102
Application programming interface
    (API), 311
Audience, 157–158
Authentication, 299
  LDAP, 300
    configure, 301, 303, 304
    enable, 300, 301
    using, 305
  OAuth
    configure, 307–309
    create, 306, 307
    using, 309, 310

B

Back up
  MySQL, 296
  PostgreSQL, 296
  SQLite, 295

Printed in the United States
by Baker & Taylor Publisher Services